CMOS Circuits Manual

CMOS Circuits Manual

R. M. Marston

Heinemann : London

William Heinemann Ltd
10 Upper Grosvenor Street, London W1X 9PA

LONDON MELBOURNE JOHANNESBURG AUCKLAND

First published 1987

British Library Cataloguing in Publication Data

Marston, R. M.
CMOS circuits manual.
1. Integrated circuits 2. Metal oxide
semiconductors
I. Title
621.3817'3 TK7874

ISBN 0 434 91212 3

Printed in England by
Redwood Burn Ltd, Trowbridge

To Kirsty, with love

Contents

Preface

One of the most important developments ever to have taken place in the digital integrated circuit field has been the introduction and development of the technology known as CMOS, COS/MOS, or COSMOS. CMOS digital ICs have considerable advantages over other digital IC types, including all members of the TTL family. In particular, they can readily be operated from unregulated supply voltages in the range 5 to 15 volts, draw virtually zero quiescent or standby current, have near-infinite input impedances, and are very easy to use. They are readily available in a large variety of device types, ranging from simple inverters and gates to complex counters and decoders, and have a multitude of practical applications in the home and the car as well as in industry.

This book is intended to act as a useful single-volume guide to the CMOS user, and is specifically aimed at the practical design engineer, technician, and experimenter, as well as the electronics student and amateur. It deals with the subject in an easy-to-read, down-to-earth, non-mathematical but very comprehensive manner. It starts off by explaining the basic principles and characteristics of the CMOS family of devices (Chapter 1), and then introduces the reader to the versatile 4007UB do-it-yourself CMOS IC, which can be used in either the digital or the linear mode (Chapter 2).

The following chapters of the volume introduce the reader to progressively more complex types of CMOS IC starting with simple inverter, gate and logic ICs and circuits, and ending with complex counters and decoders. The final chapter presents a miscellaneous collection of two dozen useful CMOS circuits.

Throughout the volume, great emphasis is placed on practical user information and circuitry, and the book abounds with useful circuits, tables, and graphs. All ICs used are inexpensive and readily available types, with universally recognized type numbers. Where possible, the outlines and pin notations of all ICs are given near the appropriate part of the text.

1 Basic principles

Digital integrated circuits (ICs) have been available for a good many years now, and most readers will be familiar with common logic family names such as TTL (transistor-transistor logic) and ECL (emitter-coupled logic), as well as those of older families such as RTL (resistor-transistor logic) and DTL (diode-transistor logic). Each of these families offers (or offered) its own particular advantages when compared with the other types, but all of them share a number of common disadvantages.

The most significant of these disadvantages are high quiescent current requirements (typically 5 mA per gate), tight power supply requirements (supplies typically have to be regulated to within 10 per cent of a specific value), low input impedance (typically a few hundred ohms per gate), and poor noise immunity (meaning that gates can easily be triggered by spikes on the input or supply lines).

In the early 1970s a new and startlingly different type of digital IC appeared on the scene, and rapidly started to push all of the older families into obsolescence in low- to medium-speed applications. This new family of devices is known as COS/MOS or CMOS (complementary-symmetry metal oxide semiconductor), and it suffers from none of the disadvantages of the earlier families.

Specifically, CMOS ICs typically draw quiescent currents of a mere 0.01 μA per gate, can be operated from any supply voltage within the range 3 to 15 volts, have a typical input impedance of about a million megohms per gate (but are fully protected against static-charge damage via built-in safety circuitry), and have inherently excellent noise immunity that enables the ICs to safely tolerate input spikes up to about 50 per cent of the supply voltage without upset.

CMOS digital ICs have excellent thermal characteristics; low-cost commercial types are designed to operate over the temperature range −40°C to +85°C, while the more expensive military versions can operate from −55°C to +125°C.

CMOS digital ICs are incredibly versatile devices; in the following chapters of this book we look at some of the many different types of IC that are now available within the CMOS family, and show how to use them. In this chapter, however, we simply explain the basic operating principles and characteristics of CMOS, and mention a few basic usage rules.

CMOS

The simplest type of circuit that can occur in any digital logic family is the inverter or NOT gate, and this element forms the basis of virtually every other type of circuit element that is used in digital electronics. *Figure 1.1(a)* shows the standard symbol of the digital inverter, and *Figure 1.1(b)* shows its RTL or resistor-transistor logic equivalent.

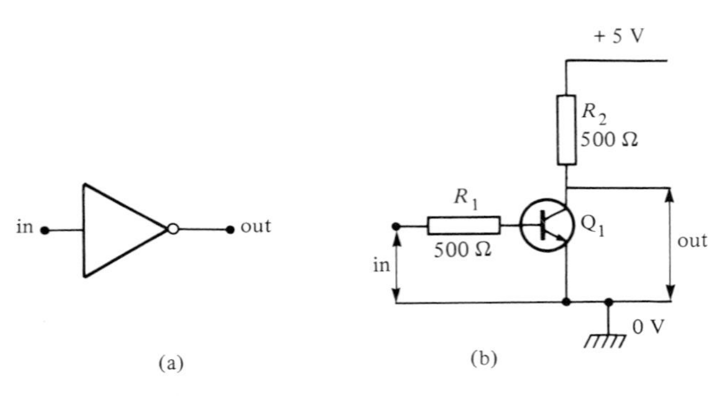

Figure 1.1 *(a) Symbol and (b) RTL equivalent of the digital inverter*

Operation of the *Figure 1.1(b)* circuit is quite simple. Inputs and outputs are always either low (grounded or at logic-0) or high (at positive supply voltage or logic-1). When the input is low, zero base drive is fed to Q_1, so Q_1 is cut off and its output is high; the quiescent current is virtually zero. When, on the other hand, the input is high, heavy base drive is applied to Q_1 via R_1, so Q_1 is driven to saturation and its output falls to the logic-0 level; the quiescent current is about 10 mA under this condition.

Thus, the RTL inverter circuit of *Figure 1.1(b)* has a fairly low input impedance (about 1.5 kΩ) and an output impedance of about 500 Ω (R_2), and draws a quiescent current of either near-zero or 10 mA. When driven by a low-frequency square wave, the circuit will draw a mean current of 5 mA. Now compare this with its CMOS equivalent.

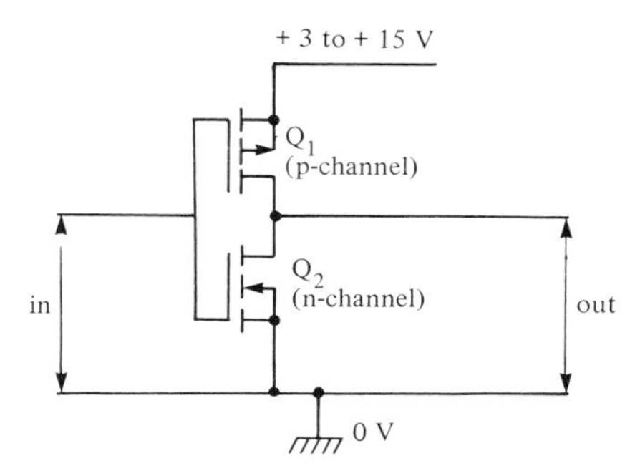

Figure 1.2 *Basic CMOS digital inverter circuit*

Figure 1.2 shows the basic circuit of a CMOS digital inverter or NOT gate. It comprises nothing more than a p-channel and an n-channel enhancement-mode MOSFET (metal-oxide semiconductor field-effect transistor) wired in series between the two supply lines, with the MOSFET gates tied together at the input terminal and with the output taken from the junction of the two devices.

The input (gate) terminal of an enhancement-mode MOSFET presents a near-infinite impedance to DC voltages, and the magnitude of an externally applied gate-to-source voltage controls the magnitude of the drain-to-source current flow. When these MOSFETs are used in the digital mode (with either a logic-0 or a logic-1 input) they can be regarded as voltage-controlled switches.

The basic digital action of the n-channel device is such that its drain-to-source path acts like an open-circuit switch when the input is at logic-0, or as a closed switch in series with a 400 Ω resistor when the input is at logic-1. The p-channel MOSFET has the inverse of these characteristics, and acts like a closed switch plus a 400 Ω resistance with a logic-0 input, and an open switch with a logic-1 input. The basic action of the CMOS digital inverter circuit can thus be understood by looking at *Figure 1.3*.

Figure 1.3(a) shows the digital equivalent of the CMOS inverter circuit with a logic-0 input. Under this condition, Q_1 (the p-channel MOSFET) acts like a closed switch in series with 400 Ω, and Q_2 acts as an open switch. The circuit thus draws zero quiescent current but can 'source' fairly large drive currents into an external output-to-ground load via the 400 Ω output resistance (R_1) of the inverter.

Figure 1.3(b) shows the equivalent of the inverter circuit with a logic-1 input. In this case Q_1 acts like an open switch, but Q_2 (the n-channel

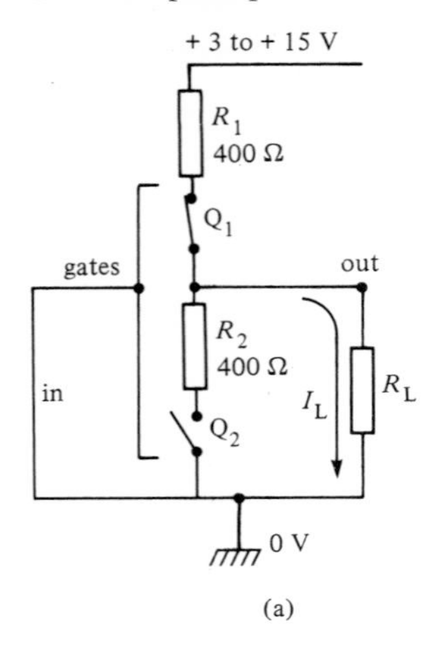

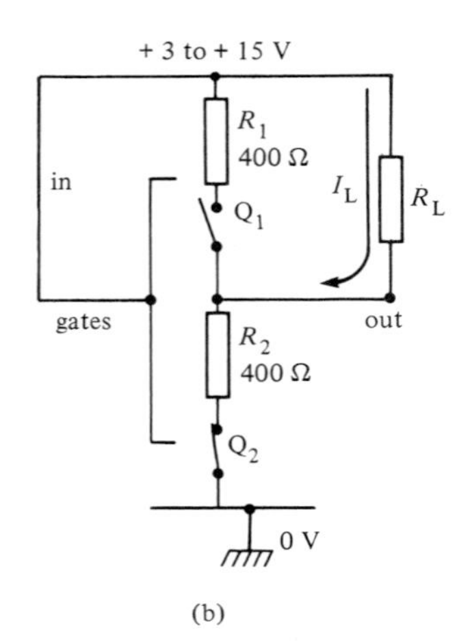

Figure 1.3 *Equivalent circuit of the CMOS digital inverter with (a) logic-0 and (b) logic-1 inputs*

MOSFET) acts like a closed switch in series with 400 Ω. The inverter thus draws zero quiescent current under this condition, but can 'sink' fairly large currents from an external supply-to-output load via its internal 400 Ω 'output' resistance (R_2).

Thus, the basic CMOS digital inverter stage has a near-infinite input impedance, draws near-zero (typically 0.01 μA) quiescent current with a logic-0 or logic-1 input, can source or sink substantial output currents, and has an output that is made inherently short-circuit proof via the 400 Ω output impedance of the device. Note that, unlike a digital inverter based on bipolar transistor technology, the output of the CMOS circuit can swing all the way from zero to the full positive supply rail voltage value, since no potentials are lost via saturation voltages or forward biased junction voltages etc. These several characteristics are in fact common to virtually all digital ICs in the CMOS range.

Linear action

All digital signals take a finite time to switch from one logic state to the other, and during this time the signal voltage can be said to have a 'linear' value. Consequently, if the reader is to fully understand CMOS circuitry, it is important that he/she should understand how the basic *Figure 1.2* CMOS inverter circuit reacts to the application of linear input voltages.

When an enhancement-mode MOSFET is used in the linear mode it acts as a voltage-controlled resistance. The n-channel device had a near-infinite (about 10000 megohms) drain-to-source resistance with zero input voltage: the resistance remains very high until the input rises to a 'threshold' value of about 1.5 to 2.5 volts, but then starts to decrease as the input voltage is increased, eventually falling to about 400 Ω when the input equals the supply line voltage. The p-channel MOSFET characteristics are the reverse of these.

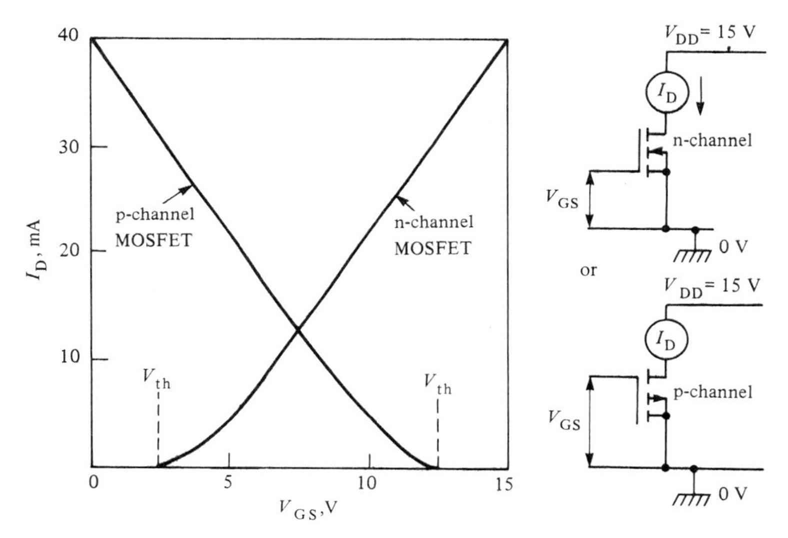

Figure 1.4 *Typical gate-voltage/drain-current characteristics of p- and n-channel MOSFETs operated from a 15 volt supply*

Thus, if either of these devices is used with a 15 volt supply, it will produce one or other of the gate-voltage/drain-current curves shown in *Figure 1.4*. Consequently, if the *Figure 1.2* CMOS inverter stage is connected to a 15 volt supply, it will produce the typical drain-current transfer graph of *Figure 1.5*, and the voltage transfer graph of *Figure 1.6*. The *Figure 1.2* circuit operation is as follows.

Suppose the input voltage is slowly increased positive, starting from zero. The inverter current is near-zero until the input exceeds the threshold voltage of the n-channel MOSFET, at which point its resistance starts to decrease and that of the p-channel MOSFET starts to increase. Under this condition the inverter current is dictated by the larger of the two resistances and is of measurable proportions. When the input is appreciably *less* than half the supply volts the n-channel MOSFET resistance is far greater than that of the p-channel device, so the output of the circuit is high or at logic-1. When the input voltage is appreciably *more* than half the supply volts the resistance of

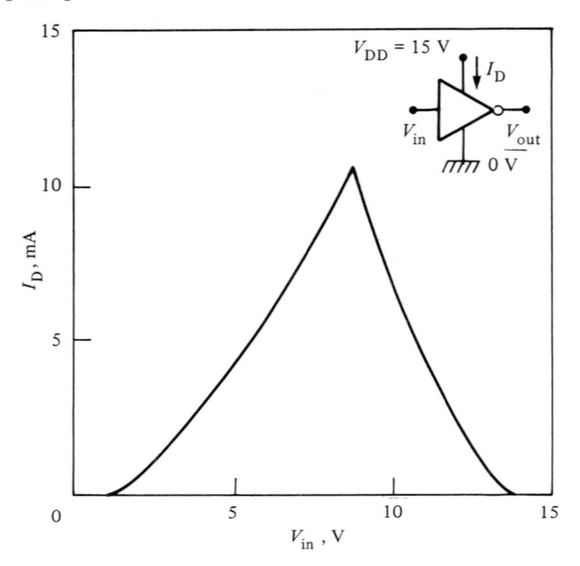

Figure 1.5 *Typical drain-current transfer characteristics of the simple CMOS inverter*

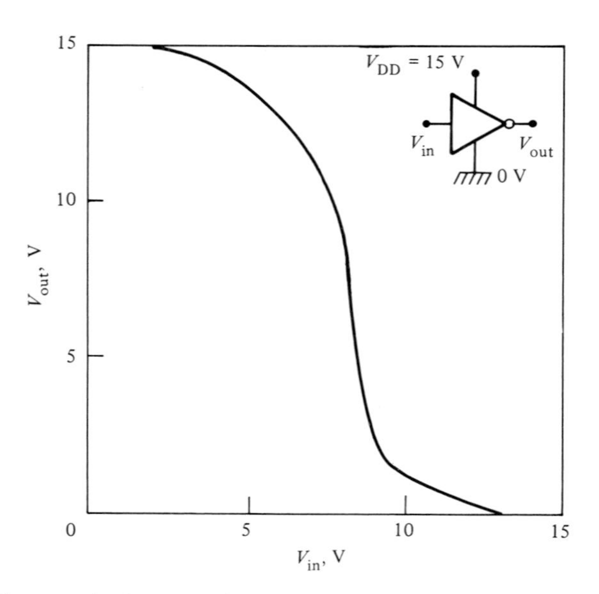

Figure 1.6 *Typical voltage transfer characteristics of the simple CMOS inverter*

the n-channel MOSFET is far less than that of the p-channel device, and the output of the circuit is thus low or at logic-0.

When the input voltage is at approximately half the supply value a point is reached where both MOSFETs have roughly the same resistance value. At this point the inverter starts to act as a linear amplifier (giving a typical voltage gain of about 30 dB), and a current of several milliamperes may flow through the circuit; small changes of input voltage cause the output to swing sharply from one logic state to the other. The value of input voltage needed to cause this 'switching' action is known as the *transition voltage*, and is specified as a percentage of the supply voltage. Transition voltages vary between 30 and 70 per cent of the supply value in practical CMOS ICs.

Finally, consider the case where the input of the *Figure 1.2* circuit is high or at logic-1. In this case, when the voltage rises to within a couple of volts of the supply value it exceeds the threshold value of the p-channel MOSFET, so the p-channel MOSFET acts like an open switch and the n-channel device acts like a 400 Ω resistance. The inverter current thus falls to near-zero under this condition.

Switching action

From the above description of CMOS linear action it should be clear that, when CMOS is driven by digital switching waveforms, it draws a brief 'pulse' of suppy current each time it goes through a switching transition. This pulse may have a peak amplitude of several milliamperes, and its duration depends (mainly) on the rise or fall time of the switching waveform. The more often CMOS changes state in a given time, the greater are the number of current pulses that it takes from the supply and the greater is its *mean* current consumption. Thus, CMOS current consumption is directly proportional to switching frequency.

At frequencies of 5 MHz, CMOS logic draws roughly the same current as its standard TTL equivalent. At 5 kHz it draws only one thousandth of the current of TTL. Consequently, CMOS is best suited to low- or medium-speed applications, although it is capable of operating as high as 10 MHz when needed.

CMOS buffering

In early CMOS logic ICs the inverter circuit took the simple form shown in *Figure 1.2*. Similarly, the two-input NOR and NAND gates took the basic forms shown in *Figures 2.18* and *2.20* respectively. These early devices were known as 'A-series' (no longer readily available) CMOS ICs, and suffered from two specific snags. The first snag is evident from the voltage transfer graph of *Figure 1.6*, which shows that (because of the inverter's low linear voltage gain) the output does not switch fully between logic states unless the input also switches fully between these states. The second snag is that A-series gates often have inherently different source and sink output impedances and current-drive values. To overcome these snags a new CMOS series was introduced, and is known as 'buffered' or 'B-series' CMOS.

A B-series CMOS IC can be simply regarded as an A-series device with one or more inverter stages wired in series with its output terminal (and sometimes also in series with its inputs). Thus, a B-series inverter can be made by wiring three A-series (*Figure 1.2*) inverters in series, as shown in *Figure 1.7*. The resulting inverter has a typical linear voltage gain of 70 to 90 dB, and gives the typical voltage transfer graph of *Figure 1.8*.

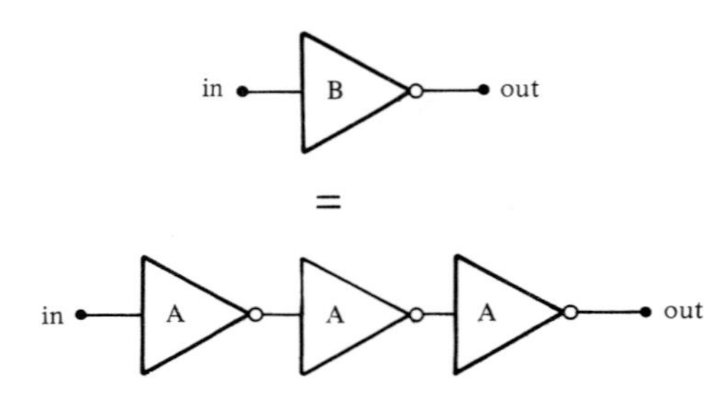

Figure 1.7 *A B-series CMOS inverter can be made by wiring three A-series types in series*

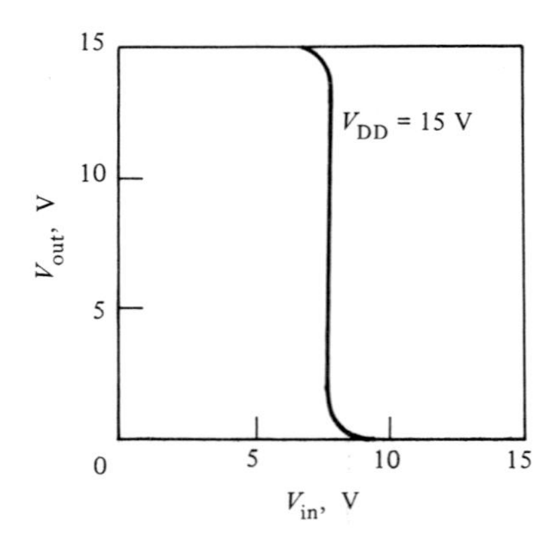

Figure 1.8 *Voltage transfer graph of the* Figure 1.7 *B-series inverter*

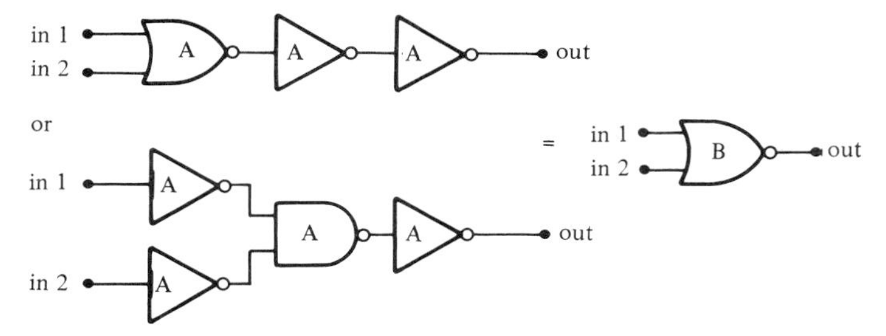

Figure 1.9 *Alternative ways of making a B-series two-input NOR gate from A-series CMOS elements*

Similarly, *Figure 1.9* shows that a B-series two-input NOR gate can be made by wiring two A-series inverters in series with the output of an A-series NOR gate (*Figure 2.18*), or by wiring an inverter in series with each input and the output of a two-input A-series NAND gate (*Figure 2.20*).

All B-series ICs carry a 'B' suffix in their type numbers (e.g. 4001B, 4093B). The serious disadvantages of B-series devices are that they have longer propagation delays and lower maximum operating frequencies than A-series devices, and are also prone to oscillation if operated in the linear mode or with slow input waveforms. Consequently, to meet special circuit design needs, a number of CMOS ICs are still produced in A-style unbuffered form; such devices carry a 'UB' suffix in their type numbers, as in the case of the 4007UB that forms the basis of Chapter 2.

Three-state outputs

The standard CMOS inverter stage produces a two-state output that (basically) is always in one or other of the low-impedance logic-0 or logic-1 states, as shown in *Figure 1.3*. Close inspection of *Figure 1.3* should make it clear, however, that a 'third' output state is also possible, and this is the one in

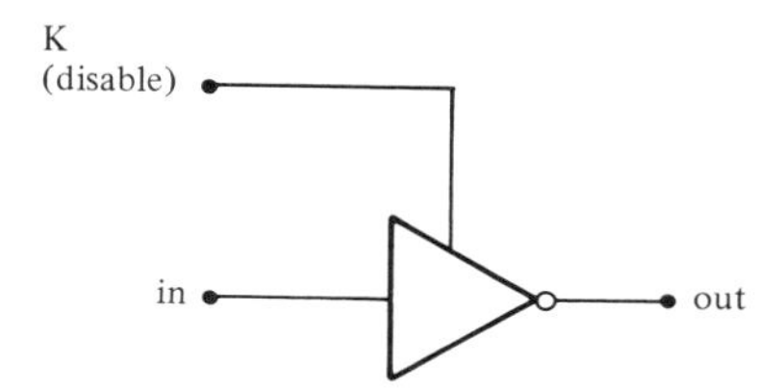

Figure 1.10 *Symbol of three-state inverter*

which Q_1 and Q_2 are both acting as open-circuit switches and the output acts as a 'floating' high impedance. Some CMOS inverters are actually made with this three-state facility built in (they sometimes form part of the output circuitry of complex ICs). *Figure 1.10* shows the standard symbol of the three-state inverter (which has an additional K (control) or 'disable' terminal that controls the third state). *Figure 1.11* shows the equivalent circuit of the inverter when the third state is operative.

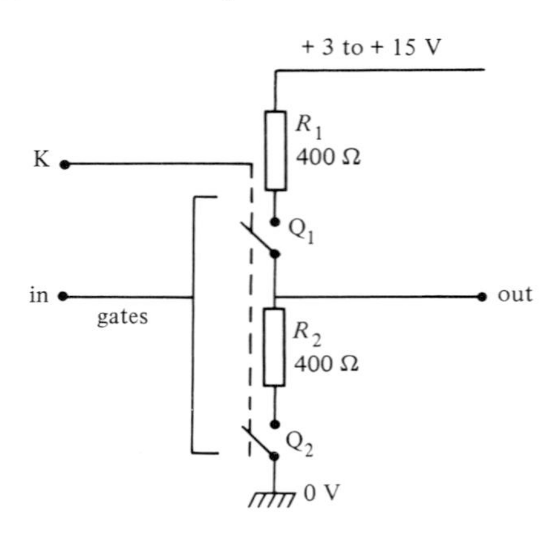

Figure 1.11 *Equivalent circuit of the three-state inverter in its 'third' state*

Input/output protection

The standard CMOS inverter circuit (on which all other CMOS digital ICs are based) has a typical input impedance of about a million megohms. To protect these inputs against the possibility of damage from static charges etc., nearly all ICs in the CMOS digital range have an integral diode-resistor protection network on every input terminal; the output terminals are provided with similar protection networks.

The precise forms of these built-in protection networks are subject to some variation, but they usually take the general form shown in *Figure 1.12*. In this network, diodes D_1 prevent the input voltage from going significantly above the positive supply rail value, D_2 stop it from going below ground value, and R_1 limits overload input currents to a few milliamperes. Similarly, D_3 and D_4 prevent the output from going above the positive supply or below the ground values, and D_5 prevents the 'positive' rail from going negative to the ground rail. R_1 and R_2 have typical values of a few hundred ohms, and all diodes have a maximum current rating of about 10 mA.

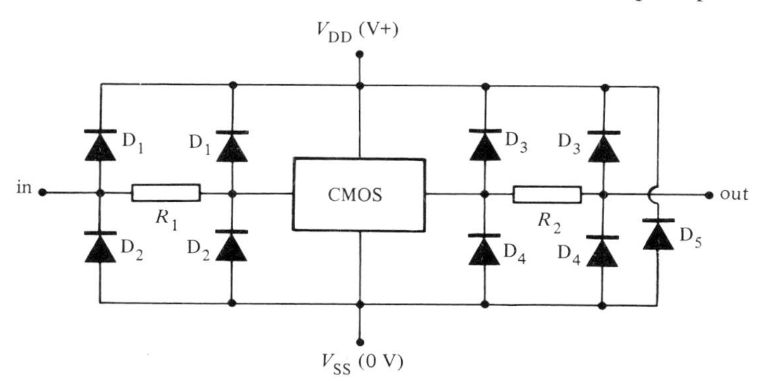

Figure 1.12 *Typical CMOS input/output protection networks*

CMOS usage rules

CMOS digital ICs are inherently very rugged devices and can withstand considerable abuse without suffering damage. Their outputs, for example, are inherently short-circuit proof, and their inputs and outputs are internally protected against excess voltages. There are in fact only three basic ways of damaging a CMOS IC (other than by using excessive supply voltages or exceeding power ratings); one of these is to connect the supply lines in the wrong polarity, in which case very heavy current will flow through D_5 of *Figure 1.12* and damage the IC substrate.

The other ways of damaging CMOS are via a *very low impedance* input or output 'signal' that is either connected to the CMOS when its power supply is switched off, or of such a large amplitude that it forces the input terminal above the positive supply line or below the zero-volts rail. In either case, a heavy current will flow through one or more of the D_1 to D_4 protection diodes, and the substrate will again be damaged. Both of these unlikely possible sources of damage can be eliminated by simply wiring a 1 kΩ resistor in series with each input/output terminal, so that any current that does flow is limited to a safe value of a few milliamperes.

Thus the basic usage rules of CMOS are quite simple. First, don't break any of the rules mentioned above. Second, always tie unused input terminals directly to either ground or the positive supply line, depending on the logic requirements. Finally, never let 'used' input terminals float; always take them to either ground or the positive line via a high-value resistor. That's all there is to it.

2 4007UB circuits

The 4007UB is the simplest IC in the entire CMOS range. It contains little more than two pairs of complementary MOSFETs, plus a simple CMOS inverter stage; all of these elements are, however, independently accessible, enabling them to be configured in a wide variety of ways, and thus making the IC the most versatile in the entire CMOS range.

The 4007UB is an ideal device for demonstrating CMOS principles to students, technicians, and engineers. It is sometimes known as the 'design-it-yourself' CMOS chip, and can readily be configured to act as a multiple digital inverter, a NAND or NOR gate, a transmission gate, or a uniquely versatile 'micropower' linear amplifier, oscillator, or multivibrator. We'll look at some practical examples of these applications later in this chapter. In the meantime, let's look at 4007UB basics.

4007UB basics

Figure 2.1(a) shows the functional diagram and pin numbering of the 4007UB, which houses two complementary pairs of independently accessible MOSFETs, plus a third complementary pair that is connected in the form of a standard CMOS inverter stage. Each of the three independent input terminals of the IC is internally connected to the standard CMOS protection network shown in *Figure 2.1(b)*. All MOSFETs in the 4007UB are enhancement-mode devices; Q_1, Q_3, and Q_5 are p-channel MOSFETs, and Q_2, Q_4, and Q_6 are n-channel MOSFETs. *Figure 2.1(c)* shows the terminal notations of the two MOSFET types; note that the B terminal represents the bulk substrate.

The term CMOS thus actually stands for complementary metal-oxide

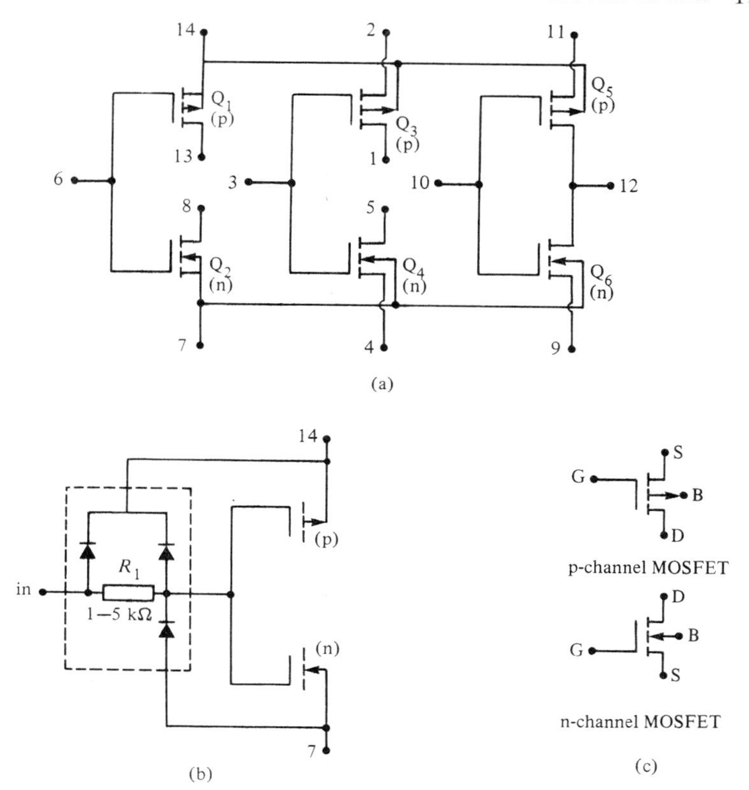

Figure 2.1 *(a) Functional diagram of the 4007UB dual CMOS pair plus inverter. (b) Internal input protection network (within dashed lines) on each input of the 4007UB. (c) MOSFET terminal notations: G=gate, D=drain S=source, B=bulk substrate*

semiconductor field-effect transistors, and it is fair to say that all CMOS ICs are designed around the basic elements shown in *Figure 2.1*. It is thus worth getting a good basic understanding of these elements. Let's look first at the digital characteristics of the basic MOSFETs.

Digital operation

The input (gate) terminal of a MOSFET presents a near-infinite impedance to DC voltages, and the magnitude of an external voltage applied to the gate controls the magnitude of source-to drain current flow. The basic characteristics of the enhancement-mode n-channel MOSFET are that the source-to-drain path is open circuit when the gate is at the same potential as the source, but becomes a near short-circuit (a low-value resistance) when the gate is heavily biased *positive* to the source. Thus the n-channel MOSFET can be used as a digital inverter by wiring it as shown in *Figure 2.2*; with a logic-0 (zero volts) input the MOSFET is cut off and the output is at logic-1 (positive rail voltage), but with a logic-1 input the output is at logic-0.

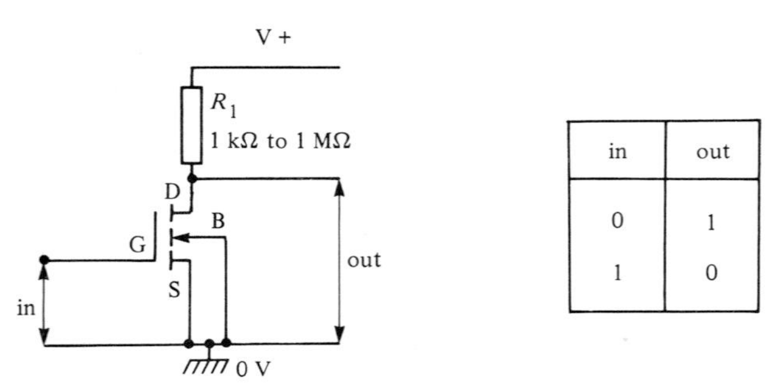

in	out
0	1
1	0

Figure 2.2 *Digital inverter made from n-channel MOSFET*

The basic characteristics of the enhancement-mode p-channel MOSFET are that the source-to-drain path is open when the gate is at the same potential as the source, but becomes a near short-circuit when the gate is heavily biased negative to the source. The p-channel MOSFET can thus be used as a digital inverter by wiring it as shown in *Figure 2.3*.

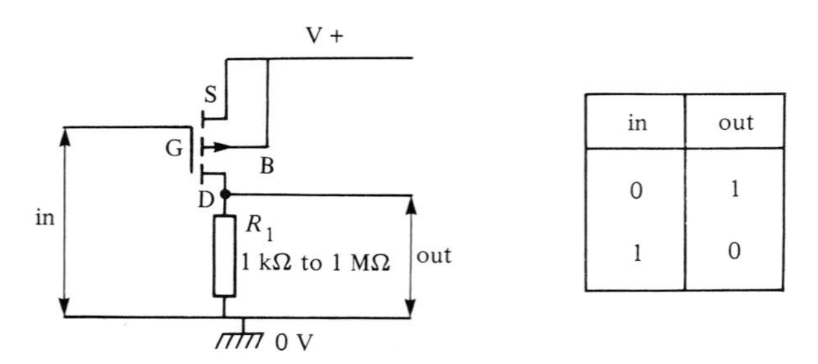

in	out
0	1
1	0

Figure 2.3 *Digital inverter made from p-channel MOSFET*

Note that in the *Figures 2.2* and *2.3* inverter circuits the on currents of the MOSFETs are determined by the value of R_1, and that these circuits draw a finite quiescent current when they are in one of their logic states. This snag can be overcome by connecting the complementary pair of MOSFETs in the classic CMOS inverter configuration shown in *Figure 2.4(a)*.

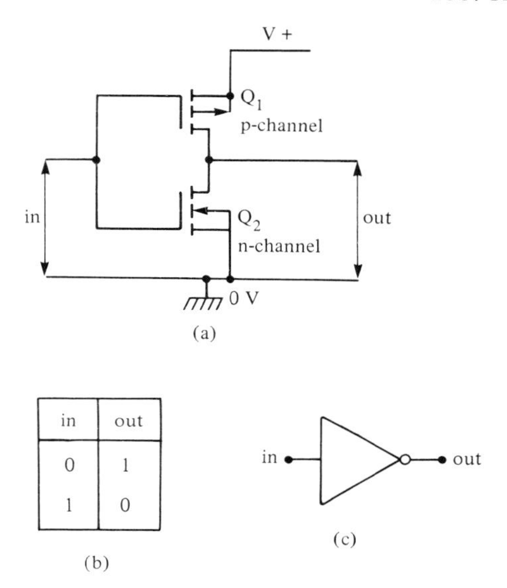

in	out
0	1
1	0

Figure 2.4 *(a) Circuits, (b) truth table and (c) standard symbol of the CMOS digital inverter*

In *Figure 2.4(a)*, with a logic-0 input applied, Q_1 is short circuited, so the output is firmly tied to the logic-1 (positive rail) state, but Q_2 is open and the inverter thus passes zero quiescent current via this transistor. With a logic-1 input applied, Q_2 is short circuited and the output is firmly tied to the logic-0 (zero volts) state, but Q_1 is open and the circuit again passes zero quiescent current.

This zero quiescent current characteristic of the complementary MOSFET inverter is one of the most important features of the CMOS range of digital ICs, and the *Figure 2.4(a)* circuit forms the basis of almost the entire CMOS family. *Figure 2.4(c)* shows the standard symbol used to represent a CMOS inverter stage. Q_5 and Q_6 of the 4007UB are fixed-wired in this inverter configuration.

Linear operation

To fully understand the operation and vagaries of CMOS circuitry, it is necessary to understand the linear characteristics of basic MOSFETs. *Figure 2.5* shows the typical gate-voltage/drain-current graph of an n-channel enhancement-mode MOSFET. Note that negligible drain current flows until the gate voltage rises to a 'threshold' value of about 1.5 to 2.5 volts, but that

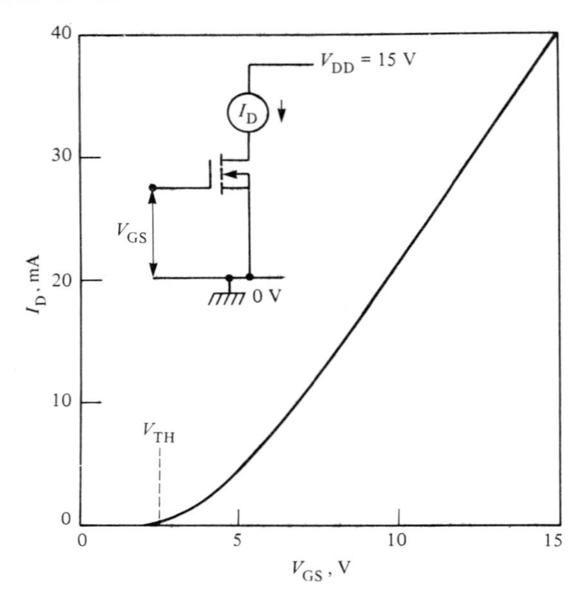

Figure 2.5 *Typical gate-voltage/drain-current characteristics of an n-channel MOSFET*

the drain current then increases almost linearly with further increases in gate voltage.

Figure 2.6(a) shows how to connect an n-channel 4007UB MOSFET as a linear inverting amplifier. R_1 serves as the drain load of Q_2, and R_2-R_x bias the gate so that the device operates in the linear mode. The R_x value must be selected to give the desired quiescent drain voltage, but is normally in the 18 to 100 kΩ range. If you want the amplifier to give a very high input impedance, wire a 10 MΩ isolating resistor between the R_2-R_x junction and the gate of Q_2, as shown in *Figure 2.6(b)*.

Figure 2.7 shows the typical I_D/V_{DS} characteristics of an n-channel MOSFET at various fixed values of gate-to-source voltage. Imagine here that, for each set of curves, V_{GS} is fixed at the V_{DD} voltage, but the V_{DS} output voltage can be varied by altering the value of drain load R_L. The graph can be divided into two characteristic regions, as indicated by the dashed line, these being the triode region and the saturated region.

When the MOSFET is in the saturated region (with V_{DS} at some value in the nominal range 50 to 100 per cent of V_{GS}) the drain acts like a constant-current source, with its current value controlled by V_{GS}; a low V_{GS} value gives a low constant-current value, and a high V_{GS} value gives a high constant-current value. These saturated constant-current characteristics provide

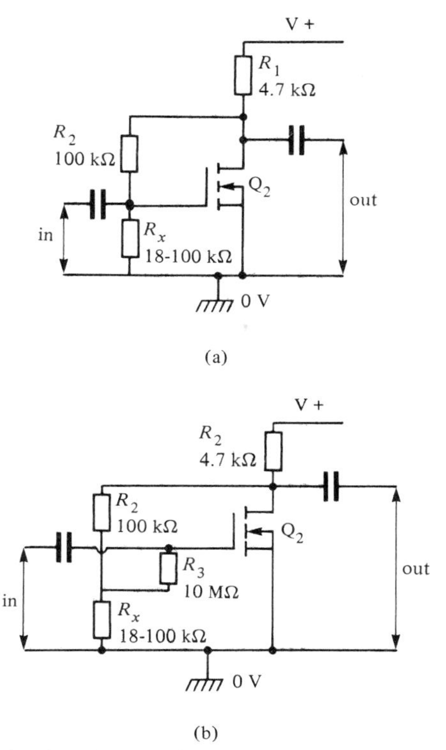

Figure 2.6 *Methods of biasing an n-channel MOSFET as a linear inverting amplifier*

CMOS with its output short-circuit proof feature and also determine its operating speed limits at different supply voltage values.

When the MOSFET is in the triode region (with V_{DS} at some value in the nominal range 1 to 50 per cent of V_{GS}) the drain acts like a voltage-controlled resistance, with the resistance value increasing approximately as the square of the V_{GS} value.

The p-channel MOSFET has an I_D/V_{DS} characteristics graph that is complementary to that of *Figure 2.7*. Consequently, the action of the standard CMOS inverter of *Figure 2.4* (which uses a complementary pair of MOSFETs) is such that its current-drive capability into an external load, and also its operating speed limits, increase in proportion to the supply rail voltage.

Figure 2.8 shows the typical voltage transfer characteristics of the standard CMOS inverter at different supply voltage values. Note that (on the 15 V V_{DD} line, for example) the output voltage changes by only a small amount when

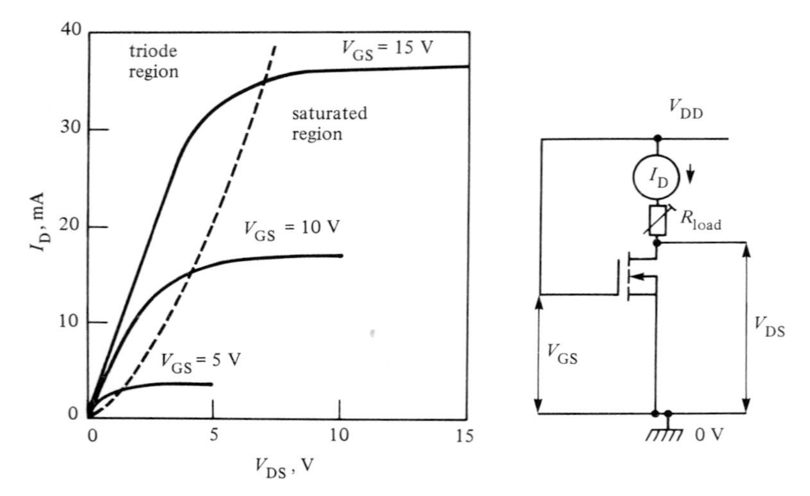

Figure 2.7 *Typical* I_D *to* V_{DS} *characteristics of the n-channel MOSFET at various fixed values of* V_{GS}

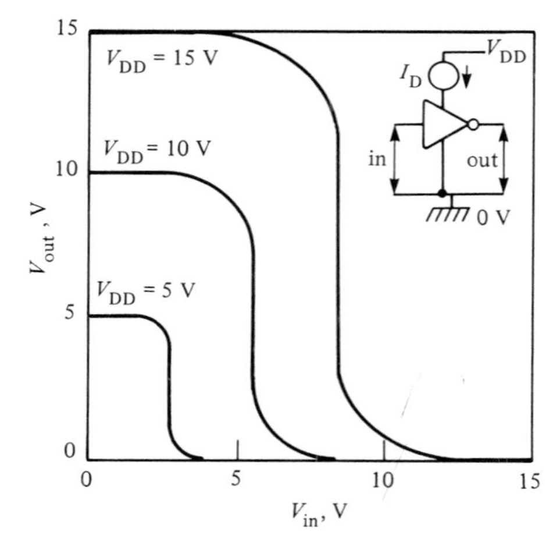

Figure 2.8 *Typical voltage transfer characteristics of the 4007UB simple CMOS inverter*

the input voltage is shifted around the V_{DD} and 0 V levels, but when V_{in} is biased at roughly half the supply volts a small change of input voltage causes a large change of output voltage: typically, the inverter gives a voltage gain of about 30 dB when used with a 15 volt supply, or 40 dB at 5 volts. *Figure 2.9*

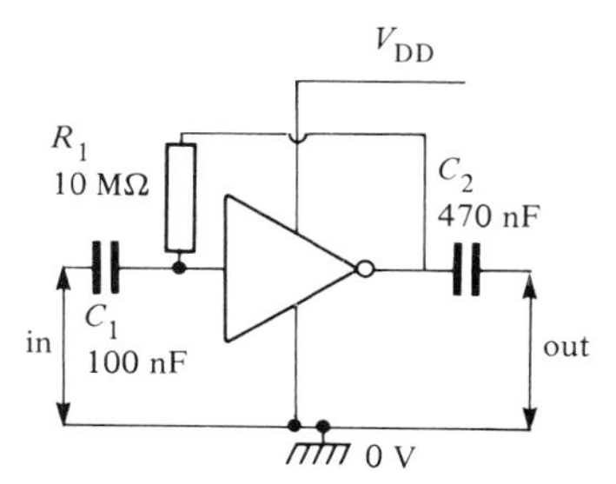

Figure 2.9 *Method of biasing the simple CMOS inverter for linear operation*

shows how to connect the CMOS inverter as a linear amplifier; the circuit has a typical bandwidth of 710 kHz at 5 volts supply, or 2.5 MHz at 15 volts.

Wiring three simple CMOS inverter stages in series (*Figure 2.10(a)*) gives the direct equivalent of a modern B-series buffered CMOS inverter stage, which has the overall voltage transfer graph of *Figure 2.10(b)*. The B-series

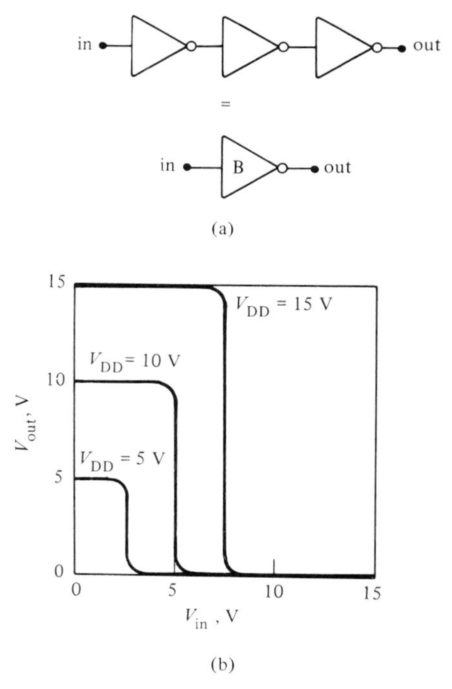

Figure 2.10 *Wiring three simple CMOS inverters in series (a) gives the equivalent of a B-series buffered CMOS inverter, which has the transfer characteristics shown in (b)*

inverter typically gives 70 dB of linear voltage gain, but tends to be grossly unstable when used in the linear mode.

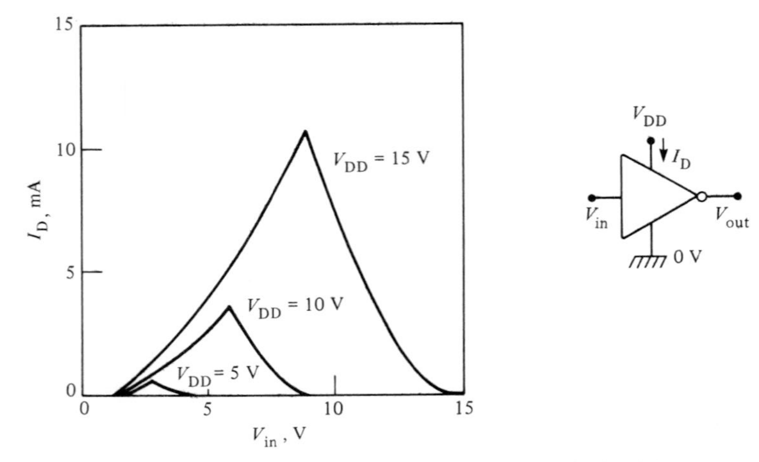

Figure 2.11 *Drain-current transfer characteristics of the simple CMOS inverter*

Finally, *Figure 2.11* shows the drain-current transfer characteristics of the simple CMOS inverter. Note that the drain current is zero when the input is at either zero or full supply volts, but rises to a maximum value (typically 0.5 mA at 5 V supply, or 10.5 mA at 15 V supply) when the input is at roughly half the supply volts, under which condition both MOSFETs of the inverter are biased on. In the 4007UB, these on currents can be reduced by wiring extra resistance in series with the source of each MOSFET of the CMOS inverter; this technique is used in the micropower circuits shown later in this chapter.

Using the 4007UB

The usage rules of the 4007UB are quite simple. In any specific application, all unused elements of the device must be disabled. Complementary pairs of MOSFETs can be disabled by connecting them as standard CMOS inverters and tying their inputs to ground, as shown in *Figure 2.12*. Individual MOSFETs can be disabled by tying their source to their substrate and leaving the drain open circuit.

In use, the input terminals must not be allowed to rise above V_{DD} (the supply voltage) or below V_{SS} (zero volts). To use an n-channel MOSFET, the source must be tied to V_{SS}, either directly or via a current-limiting resistor. To use a p-channel MOSFET, the source must be tied to V_{DD}, either directly or via a current-limiting resistor.

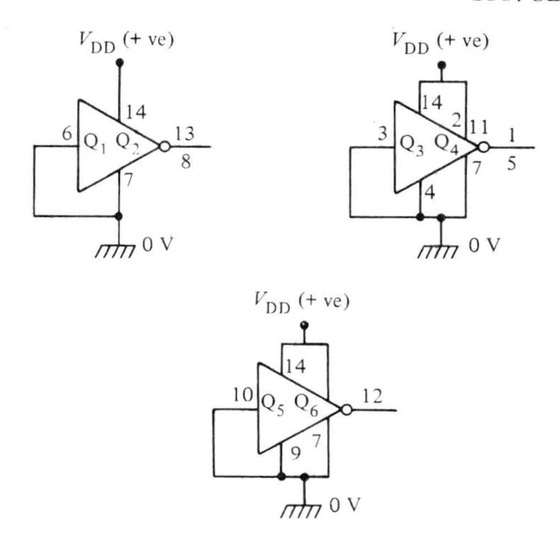

Figure 2.12 *Individual 4007UB complementary MOSFET pairs can be disabled by connecting them as CMOS inverters and grounding their inputs*

Practical circuits: digital

The 4007UB elements can be configured to act as any of a variety of standard digital circuits. *Figure 2.13* shows how to wire it as a triple inverter, using all three sets of complementary MOSFET pairs. *Figure 2.14* shows the connections for making an inverter plus non-inverting buffer; here, the Q_1-Q_2 and Q_3-Q_4 inverter stages are simply wired directly in series, to give an overall non-inverting action.

The maximum source (load-driving) and sink (load-absorbing) output currents of a simple CMOS inverter stage self-limit at 10 to 20 mA as one or other of the output MOSFETs turns fully on. Higher sink currents can be obtained by simply wiring n-channel MOSFETs in parallel in the output

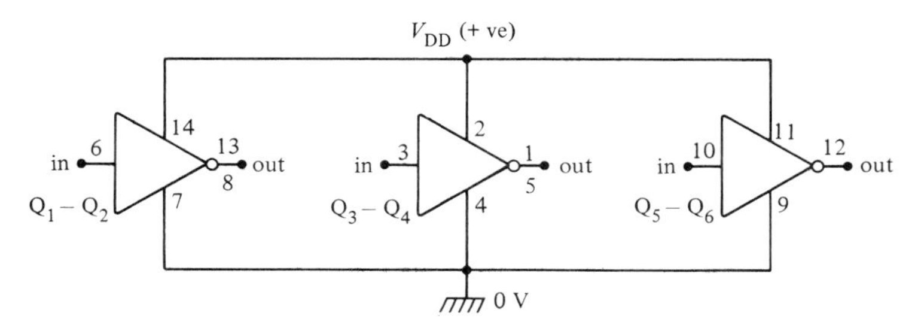

Figure 2.13 *4007UB triple inverter*

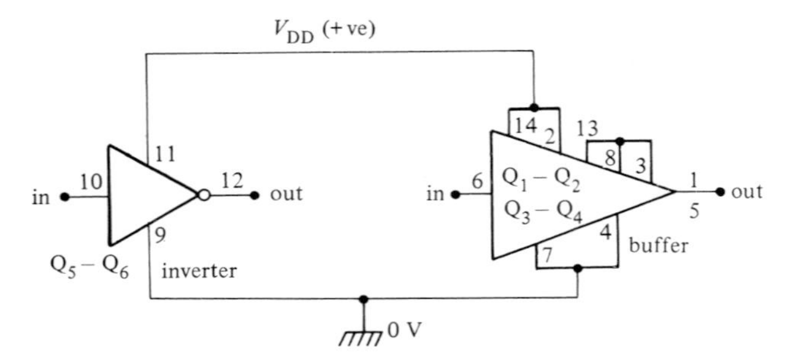

Figure 2.14 *4007UB inverter plus non-inverting buffer*

stage. *Figure 2.15* shows how to wire the 4007UB so that it acts as a high-sink-current inverter that will absorb triple the current of a normal inverter.

Similary, *Figure 2.16* shows how to wire the 4007UB to act as a high-source-current inverter, and *Figure 2.17* shows the connections for making a single inverter that will sink or source three times more current than a standard inverter stage.

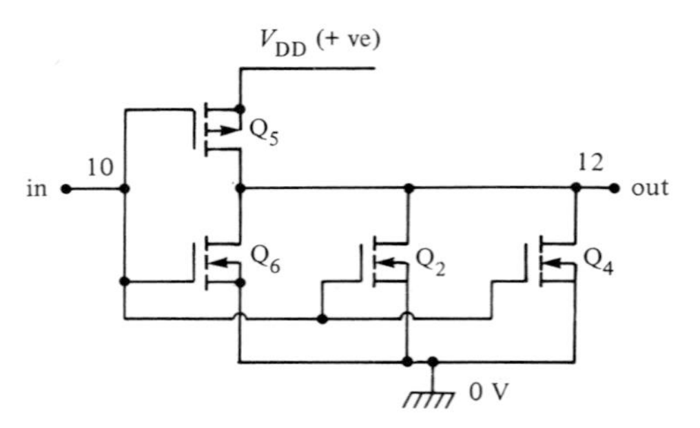

Figure 2.15 *4007UB high-sink-current inverter*

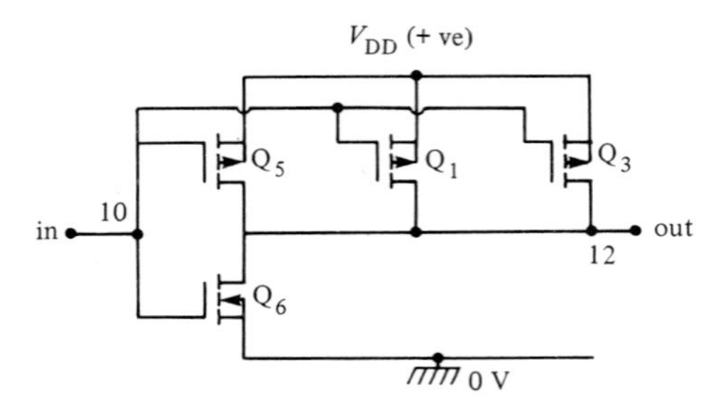

Figure 2.16 *4007UB high-source-current inverter*

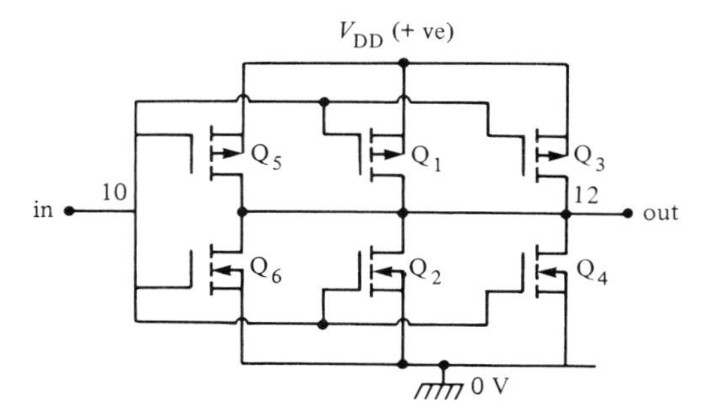

Figure 2.17 *4007UB high-power inverter, with triple the sink- and source-current capability of a standard inverter*

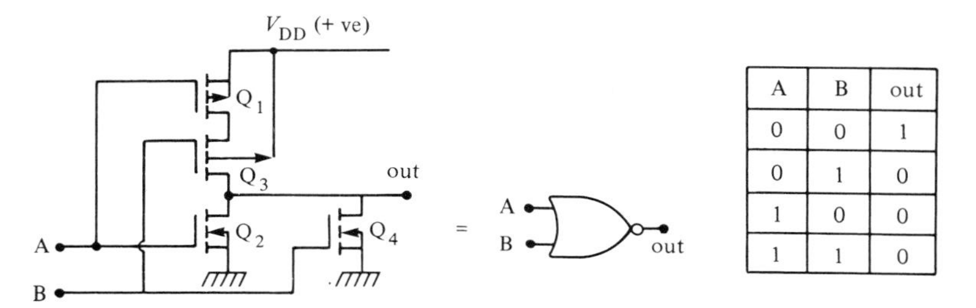

A	B	out
0	0	1
0	1	0
1	0	0
1	1	0

Figure 2.18 *4007UB two-input NOR gate*

The 4007UB is a perfect device for demonstrating the basic principles of CMOS logic gates. *Figure 2.18* shows the basic connections for making a two-input NOR gate. Note that the two n-channel MOSFETs are wired in parallel so that either can pull the output to ground from a logic-1 input, and the two

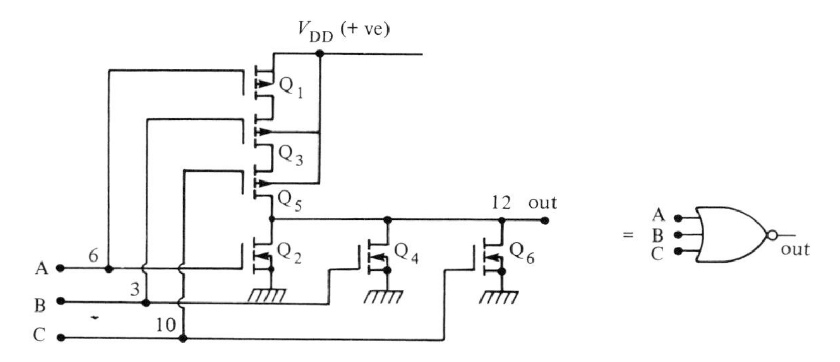

Figure 2.19 *4007UB three-input NOR gate*

p-channel MOSFETs are wired in series so that both must turn on to pull the output high from a logic-0 input. The truth table shows the logic of the circuit. A three-input NOR gate can be made by simply wiring three p-channel MOSFETs in series and three n-channel MOSFETs in parallel, as shown in *Figure 2.19*.

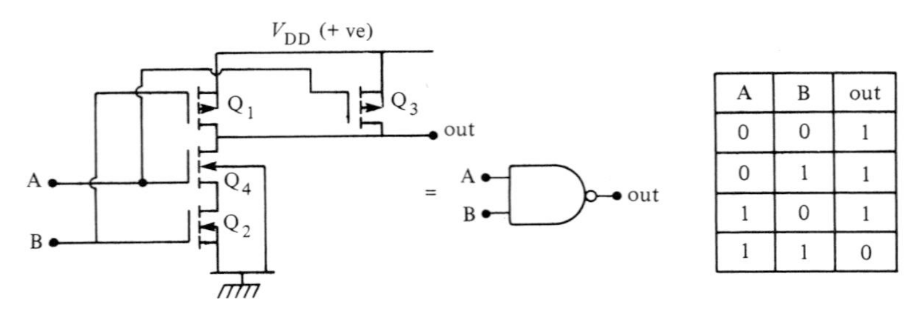

A	B	out
0	0	1
0	1	1
1	0	1
1	1	0

Figure 2.20 *4007UB two-input NAND gate*

Figure 2.20 shows how to wire the 4007UB as a two-input NAND gate. In this case the two p-channel MOSFETs are wired in parallel and the two n-channel MOSFETs are wired in series. A three-input NAND gate can be made by similarly wiring three p-channel MOSFETs in parallel and three n-channel MOSFETs in series.

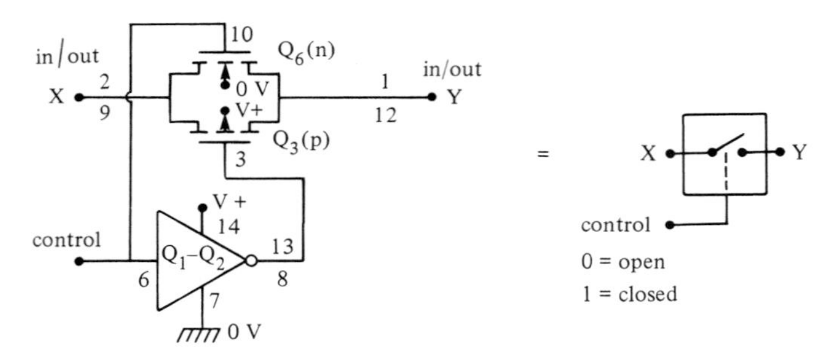

Figure 2.21 *4007UB transmission gate or bilateral switch*

Figure 2.21 shows the basic circuit for using the 4007UB to make another important CMOS element, the so-called transmission gate or bilateral switch. This device acts like a near-perfect switch that can conduct signals in either direction and can be turned on (closed) by applying a logic-1 to its control terminal or turned off (open) via a logic-0 control signal.

In *Figure 2.21* an n-channel and a p-channel MOSFET are wired in parallel (source-to-source and drain-to-drain), but their gate signals are applied in anti-phase via the Q_1-Q_2 inverter. To turn the Q_3-Q_6 transmission gate on (closed), Q_6 gate is taken to logic-1 and Q_3 gate to logic-0 via the inverter; to turn the switch off, the gate polarities are simply reversed.

The 4007UB transmission gate has a near-infinite off resistance and an on resistance of about 600 Ω. It can handle all signals between zero volts and the positive supply rail value. Note that, since the gate is bilateral, either of its terminals can function as an input or output.

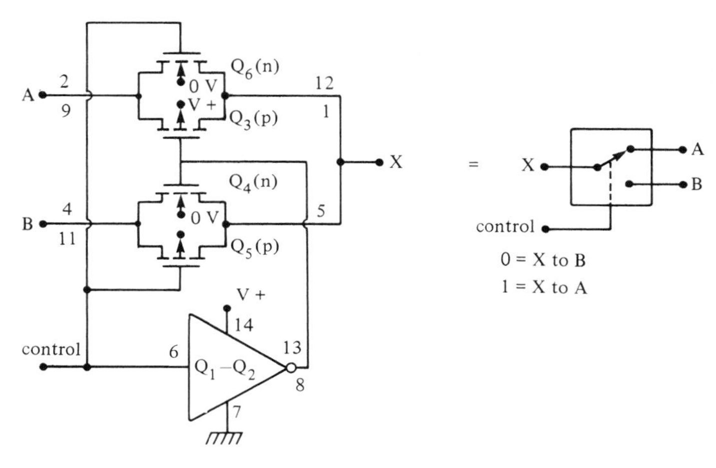

Figure 2.22 *4007UB two-way transmission gate*

Finally, *Figure 2.22* shows how the 4007UB can be wired as a dual transmission gate that functions like a single-pole double-throw (SPDT) switch. In this case the circuit uses two transmission elements, but their control voltages are applied in anti-phase, so that one switch opens when the other closes and vice versa. The X sides of the two gates are short circuited together to give the desired SPDT action.

Practical circuits: linear

We've already seen in *Figures 2.6* and *Figure 2.9* that the basic 4007UB MOSFETs and the CMOS inverter can be used as linear amplifiers. *Figure 2.23* shows the typical voltage gain and frequency characteristics of the linear CMOS inverter when operated from three alternative supply rail values (this graph assumes that the amplifier output is feeding into the high impedance of a 10 MΩ/15 pF oscilloscope probe). The output impedance of the open-loop amplifier typically varies from 3 kΩ at 15 volts supply to 5 kΩ at 10 volts or 22 kΩ at 5 volts, and it is the product of the output impedance and

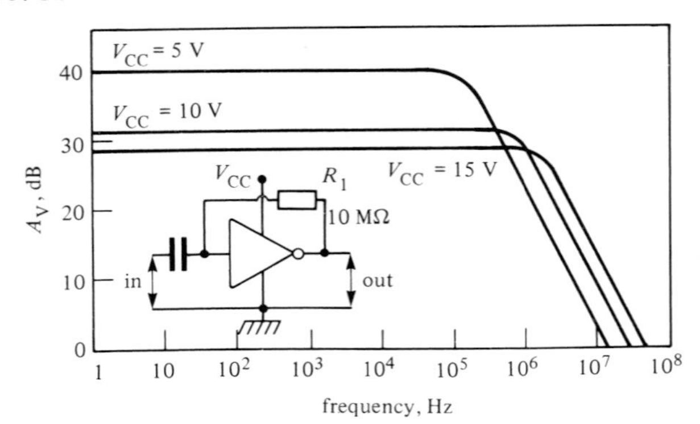

Figure 2.23 *Typical* A_V *and frequency characteristics of the linear-mode basic CMOS amplifier*

output load capacitance that determines the bandwidth of the circuit; increasing the output impedance or load capacitance reduces the bandwidth.

As you would expect from the voltage transfer graph of *Figure 2.8*, the distortion characteristics of the CMOS linear amplifier are not wonderful. Linearity is fairly good for small-amplitude signals (output amplitudes up to 3 volts peak-to-peak with a 15 V supply), but the distortion then increases progressively as the output approaches the upper and lower supply limits. Unlike a bipolar transistor circuit, the CMOS amplifier does not 'clip' excessive sine wave signals, but progressively rounds off their peaks.

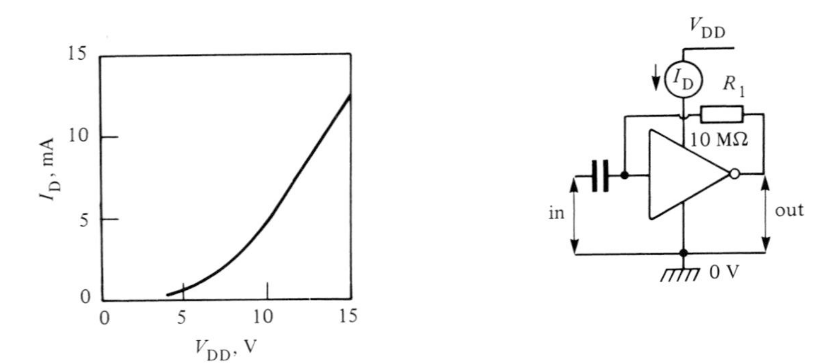

Figure 2.24 *Typical* I_D/V_{DD} *characteristics of the linear-mode CMOS amplifier*

Figure 2.24 shows the typical drain-current/supply-voltage characteristics of the basic CMOS linear amplifier. Note that the supply current typically varies from 0.5 mA at 5 volts to 12.5 mA at 15 volts.

Micropower circuits

In many applications, the quiescent supply current of the 4007UB CMOS linear amplifier can be usefully reduced, at the expense of reduced amplifier bandwidth, by wiring external resistors in series with the source terminals of the two MOSFETs of the CMOS stage, as shown in the micropower circuit of *Figure 2.25*. This diagram also shows the effect that different resistor values have on the drain current, voltage gain, and bandwidth of the amplifier when it is operated from a 15 volt supply and has its output feeding to a 10 MΩ/15 pF oscilloscope probe.

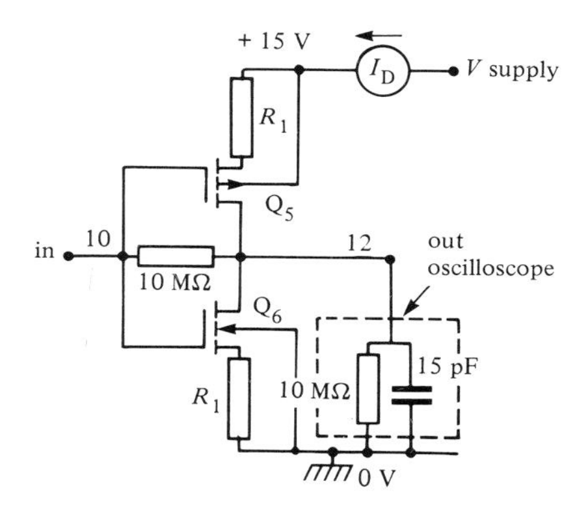

R_1	I_D	A_V (V_{out}/V_{in})	upper 3 dB bandwidth
0	12.5 mA	20	2.7 MHz
100 Ω	8.2 mA	20	1.5 MHz
560 Ω	3.9 mA	25	300 kHz
1 kΩ	2.5 mA	30	150 kHz
5.6 kΩ	600 μA	40	25 kHz
10 kΩ	370 μA	40	15 kHz
100 kΩ	40 μA	30	2 kHz
1 MΩ	4 μA	10	1 kHz

Figure 2.25 *Micropower 4007UB CMOS linear amplifier, showing method of reducing* I_D, *with measured performance details*

It is important to appreciate that in the *Figure 2.25* circuit these additional resistors add to the output impedance of the amplifier (the output impedance roughly equals the $R_1 A_V$ product), and this impedance and the external load resistance/capacitance have a great effect on the overall gain and bandwidth of the circuit. When using 10 kΩ values for R_1, for example, if the load capacitance is increased to 50 pF the bandwidth falls to about 4 kHz, but if the capacitance is reduced to a mere 5 pF the bandwidth increases to 45 kHz. Similarly, if the resistive load is reduced from 10 MΩ to 10 kΩ, the voltage gain falls to unity. Thus, for significant gain, the load resistance must be large relative to the output impedance of the amplifier.

The basic (unbiased) CMOS inverter stage has an input capacitance of about 5 pF and an input resistance of near-infinity. Thus, if the output of the *Figure 2.25* circuit is fed directly to such a load, it will show a voltage gain of about 30 and a bandwidth of 3 kHz when R_1 has a value of 1 MΩ; it will even give useful gain and bandwidth when R_1 has a value of 10 MΩ, but will consume a quiescent current of only 0.4 μA!

The CMOS linear amplifier can be used, in either its standard or its micropower forms, to make a variety of fixed-gain amplifiers, mixers, integrators, active filters and oscillators, etc. Three typical basic applications are shown in *Figure 2.26*.

A particularly attractive 4007UB linear application is as a crystal oscillator, as shown in *Figure 2.27(a)*. Here, the CMOS amplifier is linearly biased via R_1 and provides 180° phase shift, and the R_X-C_1-crystal-C_2 pi-type crystal network gives an additional 180° of phase shift at the crystal resonant frequency, thereby causing the circuit to oscillate.

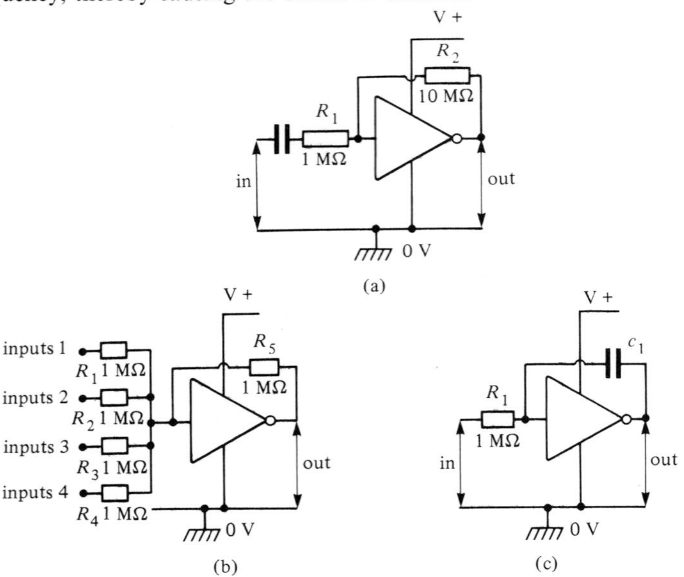

Figure 2.26 *The CMOS amplifier can be used in a variety of linear inverting amplifier applications. Three typical examples are shown here: (a) ×10 inverting amplifier (b) unity-gain four-input mixer (c) integrater*

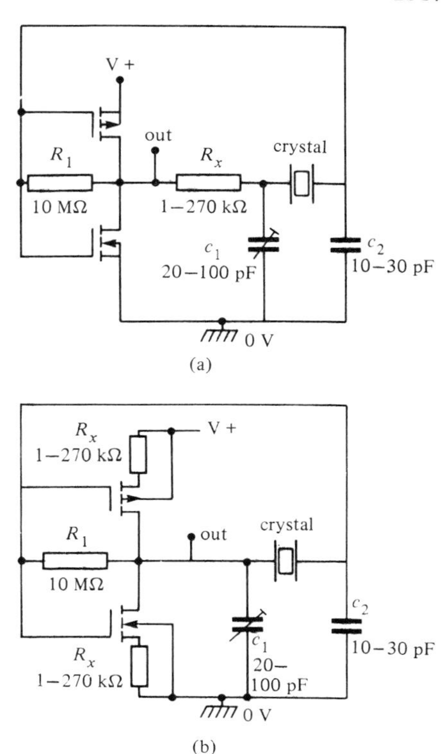

Figure 2.27 *Crystal oscillator using (a) standard and (b) micropower 4007UB CMOS linear inverter*

If in the above circuit you simply want the crystal to provide a frequency accuracy within 0.1 per cent or so, R_x can be replaced by a short-circuit and C_1-C_2 can be omitted; for ultra-high accuracy, the correct values of R_x-C_1-C_2 must be individually determined (*Figure 2.27* shows the typical range of values). In micropower applications, R_x can be incorporated in the CMOS amplifier, as shown in *Figure 2.27(b)*. If desired, the output of the crystal oscillator can be fed directly to the input of an additional CMOS inverter stage, for improved waveform shape/amplitude.

Practical circuits: astables

One of the most useful applications of the 4007UB is as a ring-of-three astable multivibrator; *Figure 2.28* shows the basic configuration of the circuit.

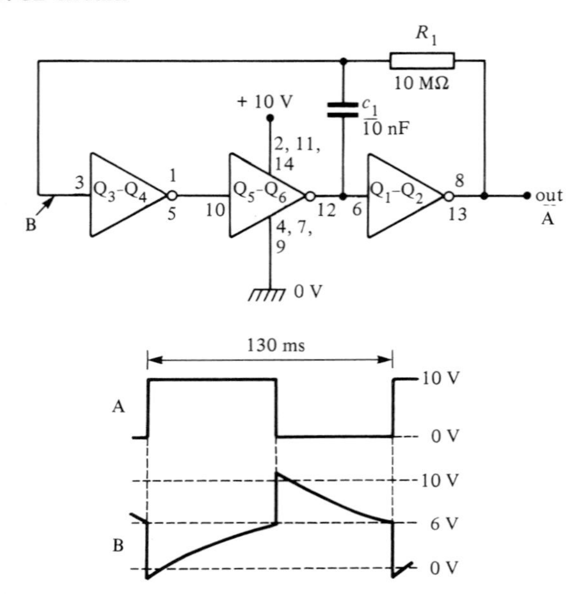

Figure 2.28 *This 4007B ring-of-three astable consumes 280 μA at 6 V, 1.6 mA at 10 V*

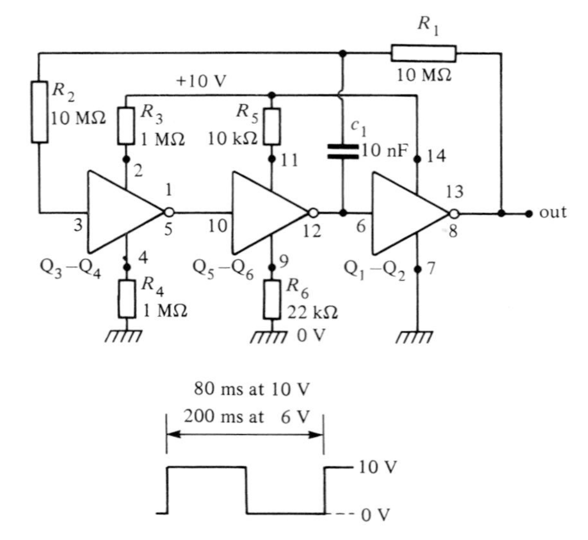

Figure 2.29 *This micropower ring-of-three symmetrical 4007UB astable consumes 1.5 μA at 6 V, 8 μA at 10 V*

Waveform timing is controlled by the values of R_1 and C_1, and the output waveform A is approximately symmetrical. Note that for most of the waveform period the front-end (waveform B) part of the circuit operates in the linear mode, so the circuit consumes a significant running current.

In practice, the running current of the *Figure 2.28* 4007UB astable circuit is higher than that of an identically configured B-series buffered CMOS IC such as the 4001B, the comparative figures being 280 μA at 6 V or 1.6 mA at 10 V for the 4007UB, against 12 μA at 6 V or 75 μA at 10 V for the 4001B. The 4007UB circuit, however, has far lower propagation delays than the 4001B and typically has a maximum astable operating speed that is three times higher than that of the 4001B.

The running current of the 4007UB astable can be greatly reduced by operating its first three stages in the micropower mode, as shown in *Figure 2.29*. This technique is of particular value in low-frequency operation, and the *Figure 2.29* circuit in fact consumes a mere 1.5 μA at 6 V or 8 μA at 10 V, these figures being far lower than those obtainable from any other IC in the CMOS range. The frequency stability of the *Figure 2.29* circuit is not, however, very good, the period varying from 200 ms at 6 V to 80 ms at 10 V.

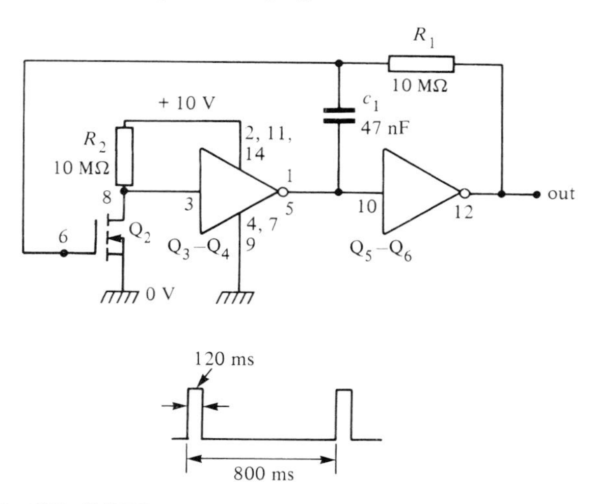

Figure 2.30 *This 4007UB asymmetrical ring-of-three astable consumes 2 μA at 6 V, 5 μA at 10 V*

Figure 2.30 shows how the 4007UB can be configured as an asymmetrical ring-of-three astable. In this case the 'input' of the circuit is applied to n-channel MOSFET Q_2. The circuit consumes a mere 2 μA at 6 V or 5 μA at 10 V.

Figure 2.31 shows how the symmetry of the above circuit can be varied by

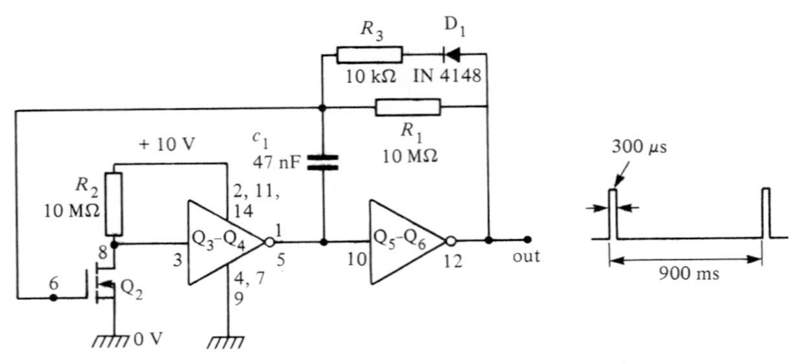

Figure 2.31 *This dual-time-constant version of the 4007UB astable generates a very narrow output pulse*

shunting R_1 with the D_1-R_3 network, so that the charge and discharge times of C_1 are independently controlled. With the component values shown, the circuit produces a 300 μs pulse once every 900 ms and consumes a mere 2 μA at 6 V or 4.5 μA at 10 V.

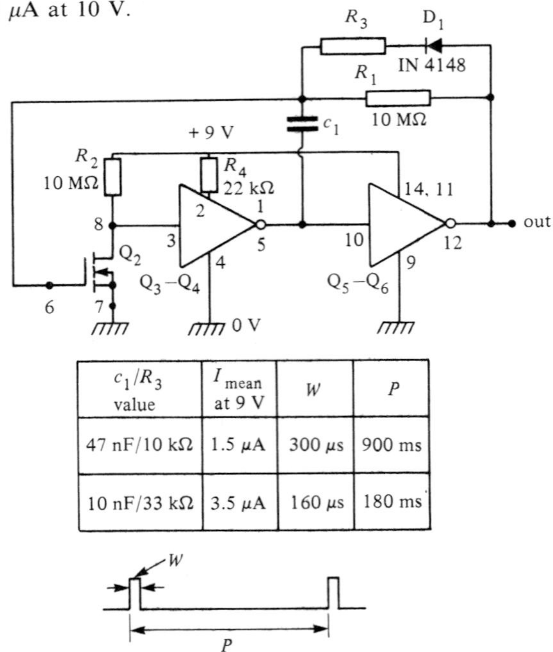

C_1/R_3 value	I_{mean} at 9 V	W	P
47 nF/10 kΩ	1.5 μA	300 μs	900 ms
10 nF/33 kΩ	3.5 μA	160 μs	180 ms

Figure 2.32 *This micropower version of the 4007UB dual-time-constant astable consumes absolutely minimal currents*

Finally, to complete this look at the 4007UB IC, *Figure 2.32* shows how the current consumption of the above circuit can be even further reduced by operating the Q_3-Q_4 CMOS inverter in the micropower mode. The table gives details of circuit performance with alternative C_1 and R_3 values. This circuit can give years of continuous operation from a single supply battery.

3 Inverter, gate, and logic circuits

Pulse inverters, buffers, and gates are the most basic elements used in digital electronics. When designing complex digital circuits, it is often necessary to work out the most economic or cost-effective way of implementing these elements. Sometimes it is best to use several discrete components (diodes, resistors, transistors, etc.) to make an element, and at others it is best to use a dedicated CMOS chip. How do you make the choice? We explain that in the next few pages.

The best known logic gates are the OR, NOR, AND, NAND, EX-OR and EX-NOR types. Less well known is 'majority' logic which, as the name implies, gives an output only when the majority of an odd number of inputs are high. Majority logic is useful in voting and pseudo-intelligent applications, such as decision-making in robotic and security systems etc. Comprehensive details of all these types of logic are given in this chapter.

Buffers and inverters

The most basic type of digital circuit is the simple pulse inverter. *Figure 3.1(a)* shows the standard circuit symbol of the inverter, and *Figure 3.1(b)* shows its truth table. *Figure 3.1(c)* shows a discrete resistor-transistor version of the inverter. In digital circuits, input and output signals are at either logic-0 (low, or zero volts) or logic-1 (high, or at full supply rail voltage) levels. Thus in *Figure 3.1(c)*, when the input is low (at logic-0) Q_1 is cut off and the output is pulled high (logic-1) via R_2, and when the input is high Q_1 is driven to saturation and its output is pulled to zero volts. The importance of the *Figure 3.1(b)* truth table is that it illustrates this information in shorthand form.

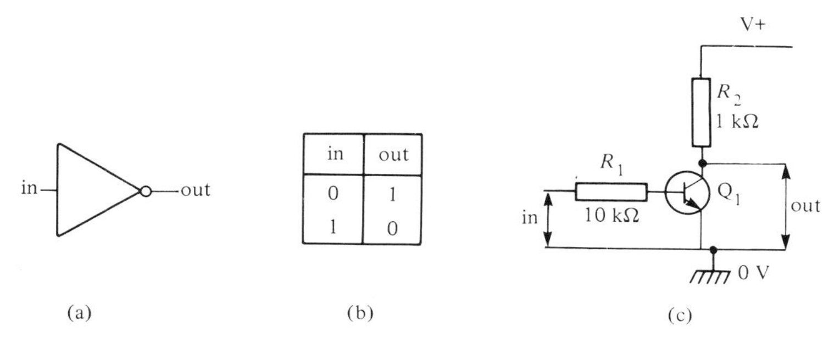

Figure 3.1 *(a) Standard symbol and (b) truth table of a digital inverter, with (c) a resistor-transistor version of the unit*

The standard inverter is the most versatile of all logic elements. It can be used to convert an OR gate into a NOR gate or vice versa, or to convert an AND gate to a NAND gate or vice versa. A pair of inverters can be used to make a bistable, monostable, or astable multivibrator, etc.

Usually, a practical inverter has an input impedance that is high relative to its output impedance, and can be used as an impedance 'buffer'. Not all buffers are of the inverting type, and *Figure 3.2(a)* shows the standard circuit symbol of a non-inverting buffer stage, which can be made by cascading two inverting elements as shown in *Figure 3.2(c)*.

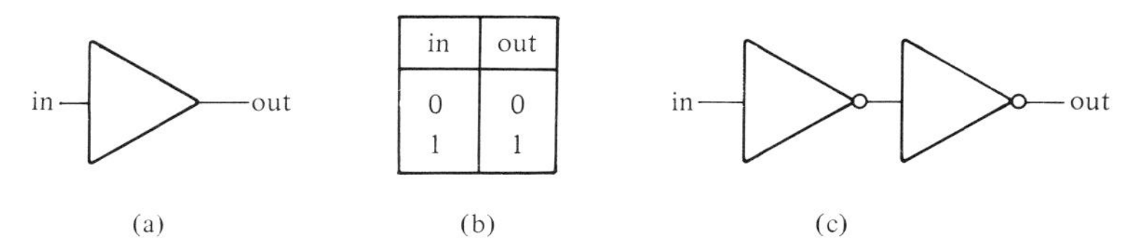

Figure 3.2 *(a) Symbol and (b) truth table of a non-inverting buffer stage, which can be made by (c) cascading two inverter stages*

Inverters and buffers are readily available in dedicated CMOS IC form, and *Figure 3.3* gives details of five popular examples. The 4041, 4049, and 4069 types use the unbuffered (UB) low-gain CMOS construction form, and the 4050 and 4502 use the high-gain buffered CMOS construction form.

The 4069UB is a simple general-purpose hex (six-element) inverter, housed in a 14-pin package, and has 'standard' output drive capability. The 4049UB hex inverting buffer and the 4050B hex non-inverting buffer, on the other hand, have high output drive capability and are specifically intended to drive TTL loads; they can accept input signal levels far higher than the supply voltage, and so can be used to give signal-level translation between CMOS and TTL circuits.

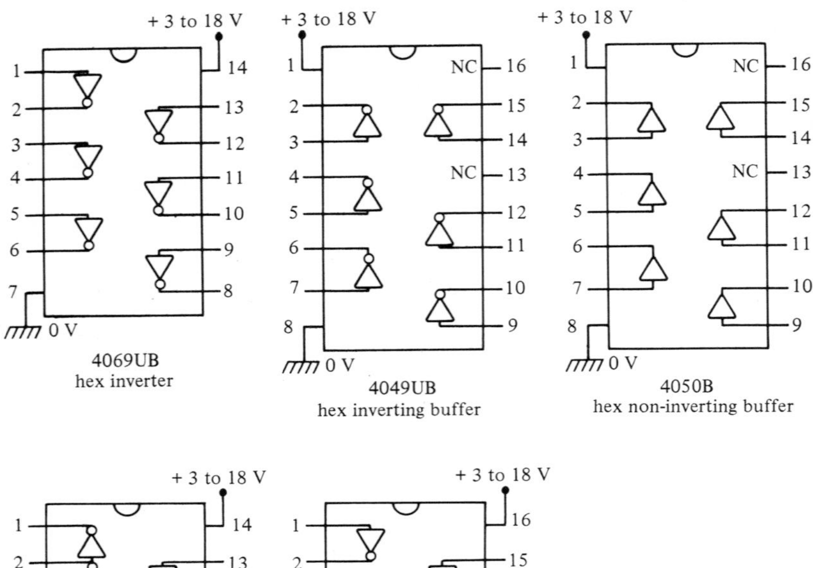

Figure 3.3 *Five popular CMOS inverter and buffer ICs*

The 4041UB also has a high output drive capability and can be used to drive TTL, but it cannot accept inputs greater than its supply voltage. The device is a quad inverting/non-inverting buffer. If, for example, an input is applied at pin 3, an inverted output is available at pin 2 and a non-inverted output at pin 1.

The 4502B is a hex inverting buffer capable of driving TTL loads, and has a tristate output that can be selected via pin 4. When pin 4 is low the IC gives 'normal' inverting operation, but when pin 4 is high all outputs go into the high-impedance tristate mode. The IC also has an 'inhibit' control terminal (pin 12), which is normally held low but which drives all outputs to ground (in the normal mode) when pin 12 is taken high.

The basic guidance rules for using inverters and buffers in practical circuits are quite simple. If you need a large number of stages, use as many dedicated ICs as necessary. If you get to a point where you are short of just one or two stages, see if you can make them from spare stages of existing logic ICs (we show how later in this chapter) or, failing that, consider using simple resistor-transistor stages of the *Figure 3.1(c)* type.

OR and NOR gates

Figure 3.4(a) shows the standard symbol of a two-input OR gate, and *Figure 3.4(b)* shows its truth table. As implied by its name, the output of an OR gate

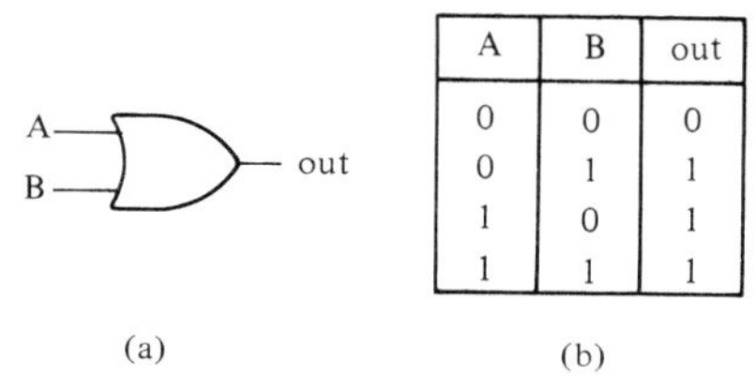

A	B	out
0	0	0
0	1	1
1	0	1
1	1	1

Figure 3.4 *(a) Symbol and (b) truth table of a two-input OR gate*

goes high if any of its inputs (A OR B etc.) go high. The simplest way to make an OR gate is to use a number of diodes and a single load resistor, as shown in the three-input OR gate of *Figure 3.5*. The diode OR gate is reasonably fast, very cost effective, and can readily be expanded to accept any number of inputs by simply adding one more diode to the circuit for each new input.

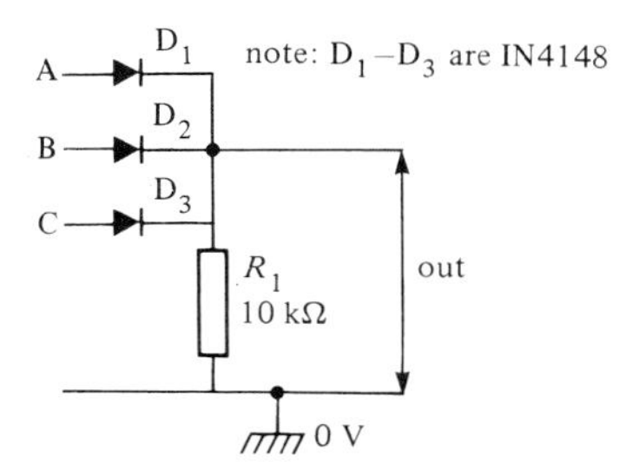

Figure 3.5 *Three-input diode OR gate*

Figure 3.6(a) shows the standard symbol of a two-input NOR gate (which functions like an OR gate with an inverted output) and *Figure 3.6(b)* shows its truth table.

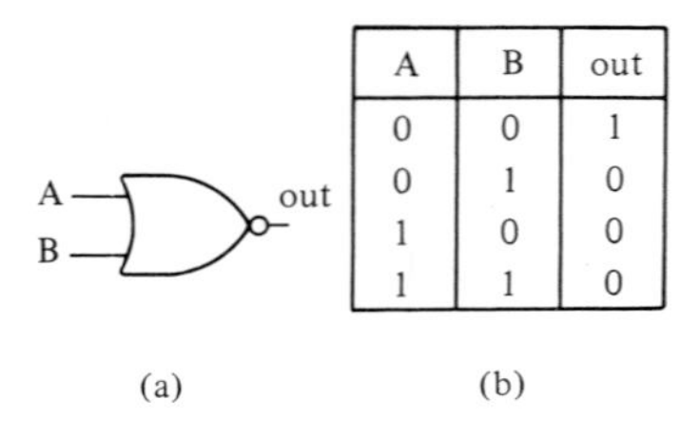

A	B	out
0	0	1
0	1	0
1	0	0
1	1	0

Figure 3.6 *(a) Symbol and (b) truth table of a two-input NOR gate*

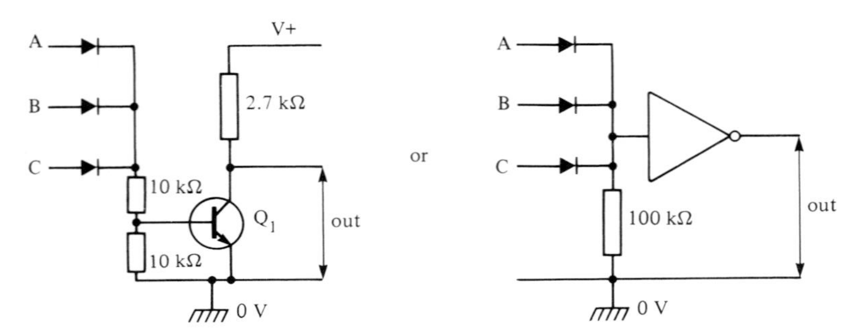

Figure 3.7 *The diode OR gate can be converted to a NOR type by feeding its output through a transistor or IC inverter*

Figure 3.7 shows how a diode OR gate can be converted to a NOR type by feeding its output through a transistor or IC inverter stage. *Figure 3.8* drives this lesson home by pointing out that an OR gate can be made from a NOR gate plus an inverter, or a NOR gate can be made from an OR gate plus an inverter.

Figure 3.8 *An OR gate can be made from a NOR gate, or vice versa, by taking the output via an inverter*

Figure 3.9 shows that a NOR gate can be made to act as a standard inverter, and an OR gate can be made to act as a non-inverting buffer, either by grounding all but one of its inputs or by connecting all inputs in parallel.

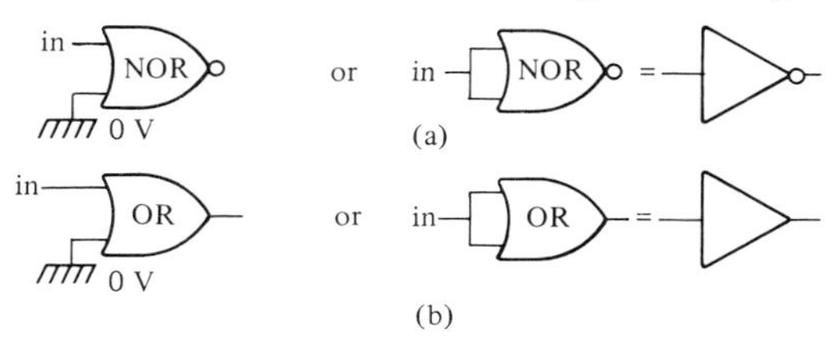

Figure 3.9 *A NOR gate can be converted to an inverter, and an OR gate can be converted to a non-inverting buffer*

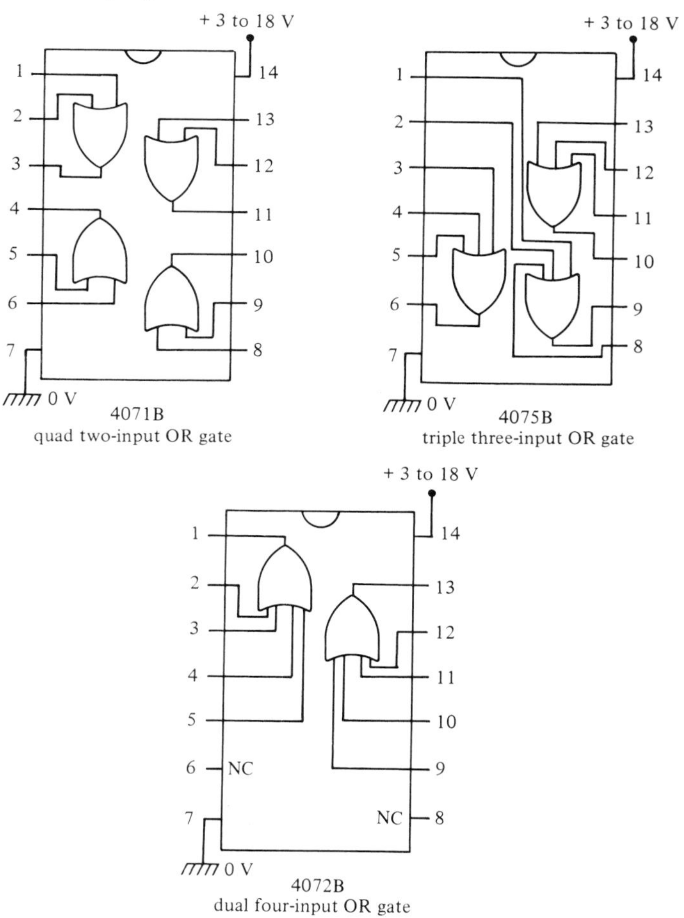

Figure 3.10 *Three popular CMOS OR gate ICs*

Figure 3.10 gives details of three popular CMOS OR gate ICs, the 4071B quad two-input type, the 4075B triple three-input type, and the 4072B dual four-input type. When using IC OR gates, note (*Figure 3.11(a)*) that the effective number of inputs can be reduced by grounding all unwanted inputs, or can be increased (*Figures 3.11(b)* and *3.11(c)*) by adding more OR gates (either integrated or discrete) to one of the inputs.

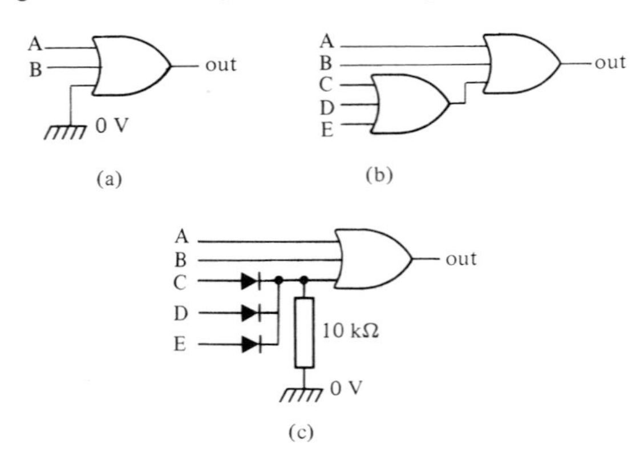

Figure 3.11 *The effective number of inputs of a CMOS OR gate can be (a) reduced by grounding all unwanted inputs, or (b), (c) increased by adding more OR gates to one of the OR inputs*

Figure 3.12 gives details of five popular CMOS NOR gate ICs. The 4001B, 4025B, and 4002B are quad two-input, triple three-input, and dual four-input devices, respectively. The 4000B contains two three-input NOR gates and a single inverter, and the 4078B is an eight-input gate that gives an OR output at pin 1 and a NOR output at pin 13.

Note that, since a NOR gate is equal to an OR gate with an inverted output, the effective number of inputs of a NOR gate can be increased or reduced by using the techniques that have already been shown in *Figure 3.11.*

A design example

Figure 3.13 illustrates a simple example of logic design using OR and NOR gates and inverters, the aim being to design a simple low-powered tone generator (driving a PB-2720 or similar acoustic transducer) that can be activated via any one of four inputs. Look first at *Figure 3.13(a)*. At first sight, the design seems to call for the use of a four-input OR gate, with its output feeding to a gated tone generator.

A suitable tone generator can be made by connecting a two-input NOR

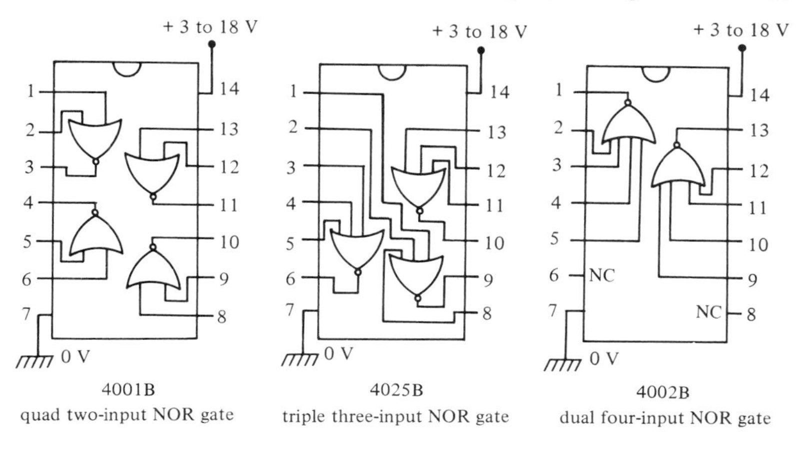

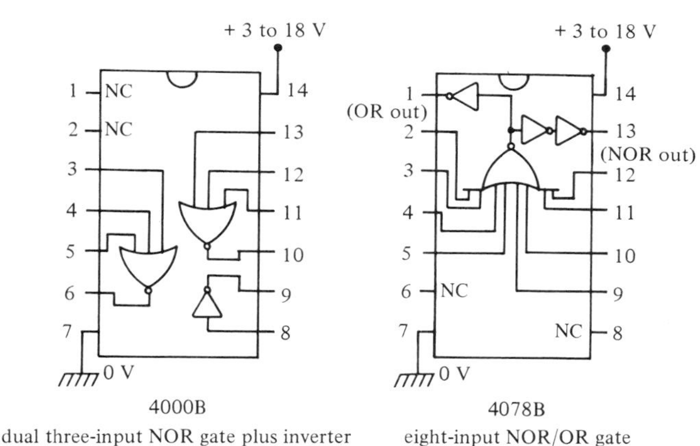

Figure 3.12 *Popular CMOS NOR gate ICs*

gate and an inverter in the standard astable configuration shown. However, this astable is gated on by low input signals, so (in *Figure 3.13(a)*) the required circuit action can be obtained by interposing an inverting stage between the output of the four-input OR gate and the input of the astable. The *Figure 3.13(a)* design thus calls for the use of three ICs.

Figure 3.13(b) shows a simple rationalization of the *Figure 3.13(a)* circuit which enables the IC count to be reduced to two. Here, the four-input OR gate plus inverter of *Figure 3.13(a)* is replaced by a four-input NOR gate, and the inverter section of the astable is made from a two-input NOR gate with its inputs short circuited together.

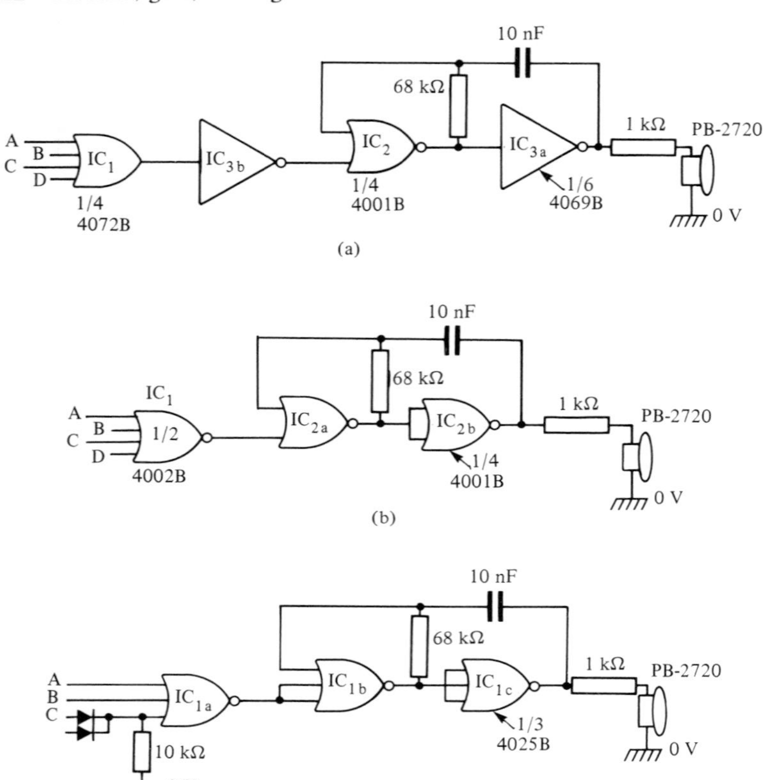

Figure 3.13 *Low-power tone generator activated by any one of four 'high' inputs. The 'over-designed' version shown in (a) uses three CMOS ICs, but the rationalized design shown in (b) uses only two CMOS chips. In (c) the design is further rationalized so that it uses only a single IC*

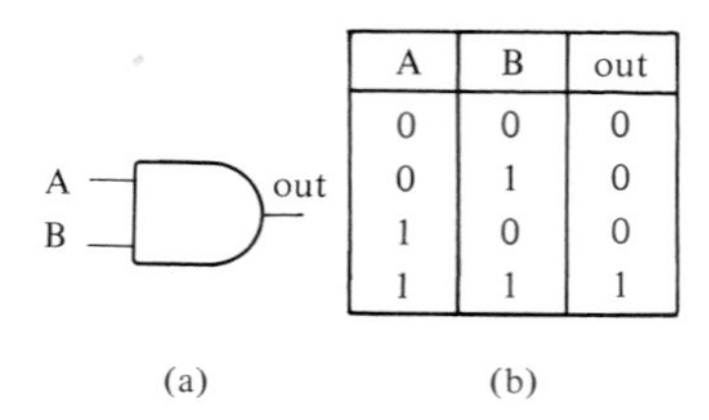

A	B	out
0	0	0
0	1	0
1	0	0
1	1	1

Figure 3.14 *(a) Symbol and (b) truth table of a two-input AND gate*

Finally, *Figure 3.13(c)* shows how the design can be further rationalized so that it uses only a single IC (a triple three-input NOR gate) and a couple of diodes. Here, the astable is made by converting a three-input NOR gate into a two-input type by short circuiting two of its inputs together, and by short circuiting all three inputs of another gate together to make an inverter, and the input gate of the circuit is converted to a four-input type by connecting a two-input diode OR gate to one of its inputs.

AND and NAND gates

Figure 3.14 shows the standard symbol and truth table of a two-input AND gate which, as indicated by its name, gives a high output when all of its inputs (A AND B etc.) go high. The simplest way to make an AND gate is to use a number of diodes and a single load resitor, as shown in the three-input AND gate of *Figure 3.15*; more inputs can be obtained by simply adding one extra diode for each new input.

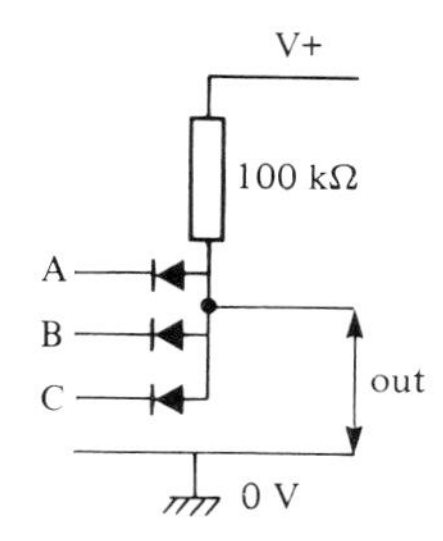

Figure 3.15 *Three-input diode AND gate*

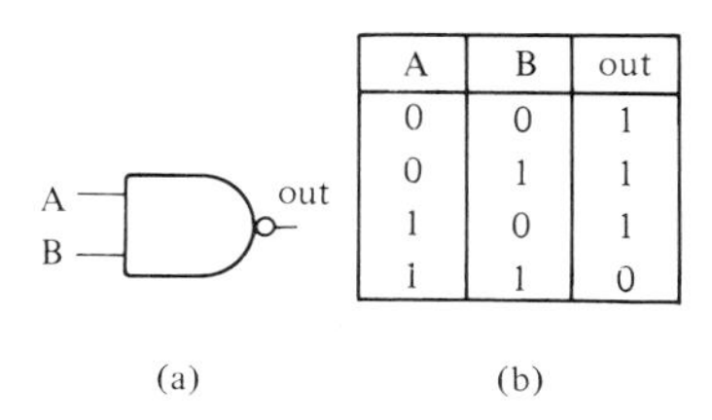

A	B	out
0	0	1
0	1	1
1	0	1
1	1	0

Figure 3.16 *(a) Symbol and (b) truth table of a two-input NAND gate*

Figure 3.16(a) shows the standard symbol of a two-input NAND gate (which functions like an AND gate with an inverted output) and *Figure 3.16(b)* shows its truth table. *Figure 3.17* shows how a NAND gate can be made from an AND gate and an inverter, and an AND gate can be made from a NAND gate and an inverter.

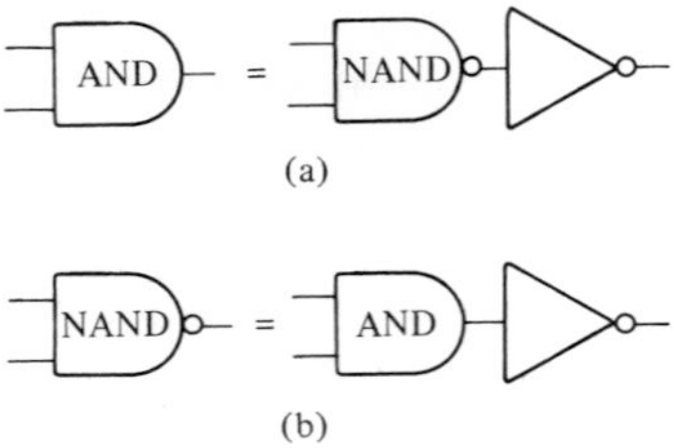

Figure 3.17 *An AND gate can be made from a NAND gate, or vice versa, by taking the output via an inverter*

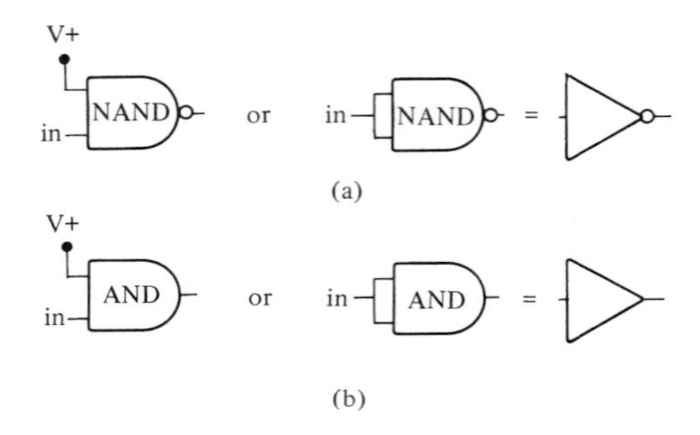

Figure 3.18 *A NAND gate can be made to act as an inverter, and an AND gate can be made to act as a non-inverting buffer*

Figure 3.18 shows that a NAND gate can be made to act as an inverter and an AND gate can be made to act as a non-inverting buffer, either by wiring all but one of the inputs to the positive (logic-1) rail or by wiring all inputs in parallel.

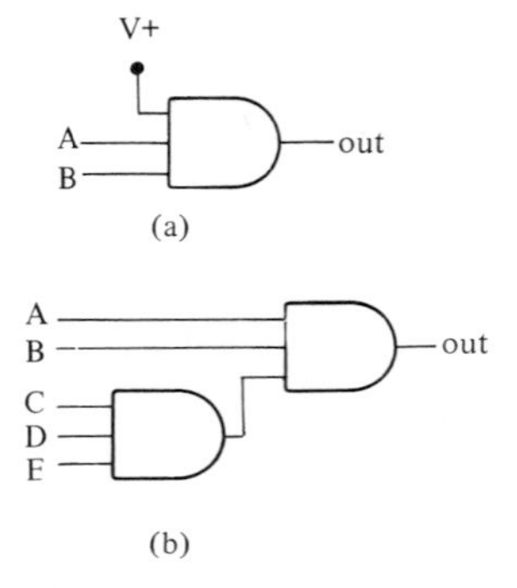

Figure 3.19 *The effective number of inputs of an AND or NAND gate can easily be (a) reduced or (b) increased*

Figure 3.19 shows that the effective number of inputs of an AND or NAND gate can be (a) reduced by wiring all unwanted inputs to the positive supply rail, or (b) increased by wiring extra AND gates to one of the inputs.

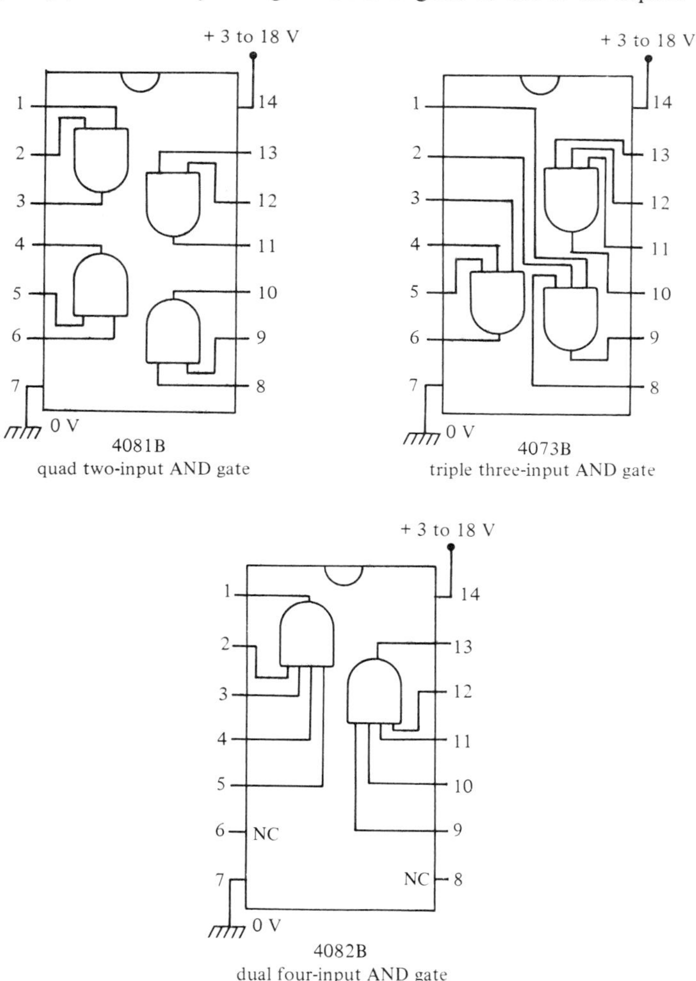

Figure 3.20 *Three popular CMOS AND gate ICs*

Figure 3.20 gives details of three popular CMOS AND gates, the 4081B quad two-input type, the 4073B triple three-input type, and the 4082B dual four-input type.

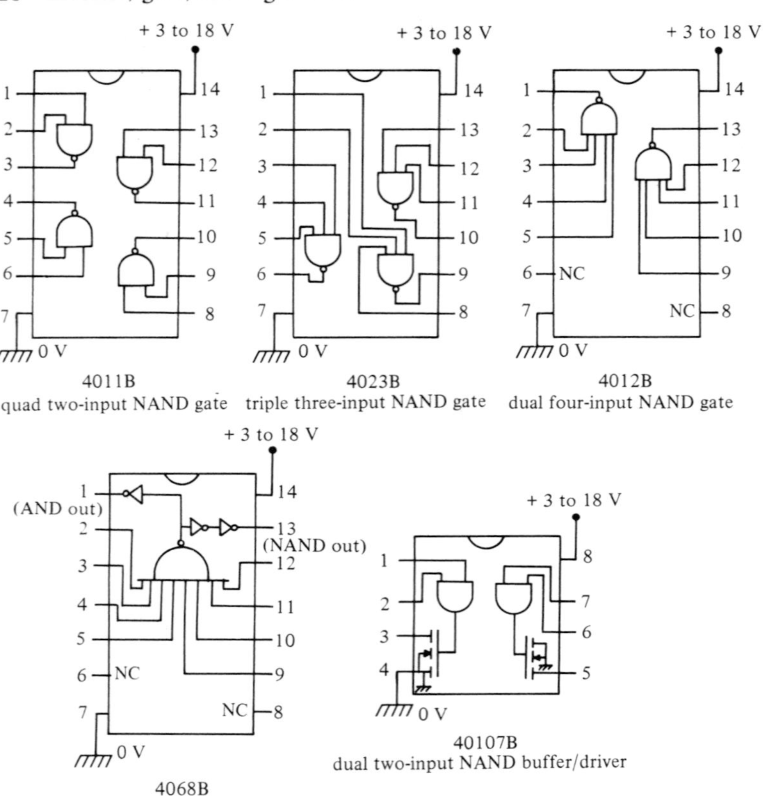

Figure 3.21 *Five popular CMOS NAND gate ICs*

Figure 3.21 gives details of five popular CMOS NAND gates. The 4011B, 4023B, and 4012B are quad two-input, triple three-input and dual four-input types, respectively. The 4068B is an eight-input device with both AND and NAND outputs. The 40107B is a dual two-input NAND gate, housed in an eight-pin package, with outputs via open-drain n-channel MOSFETs that can sink about 136 mA.

EX-OR and EX-NOR gates

Figure 3.22(a) shows the standard symbol of a two-input EX-OR (exclusive-OR) gate, and *Figure 3.22(b)* shows its truth table. The ouput of the EX-OR gate goes high only when the two inputs differ. A useful feature of the EX-OR

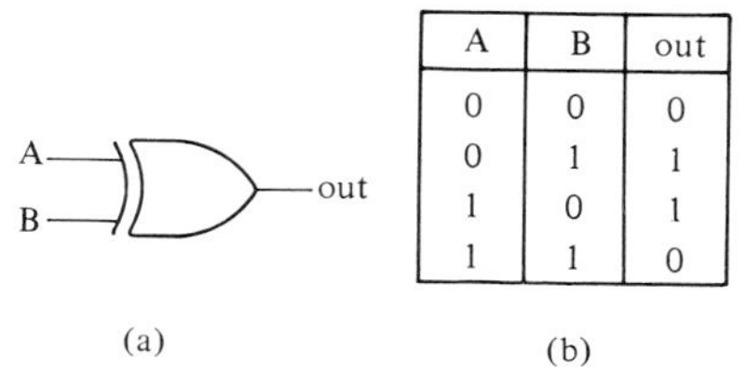

A	B	out
0	0	0
0	1	1
1	0	1
1	1	0

Figure 3.22 *(a) Symbol and (B) truth table of a two-input EX-OR gate*

gate is that it can be used as either an inverting or a non-inverting amplifier by wiring or switching one of its inputs either to the positive (logic-1) supply rail (inverting mode) or to ground (non-inverting mode), as shown in *Figure 3.23*.

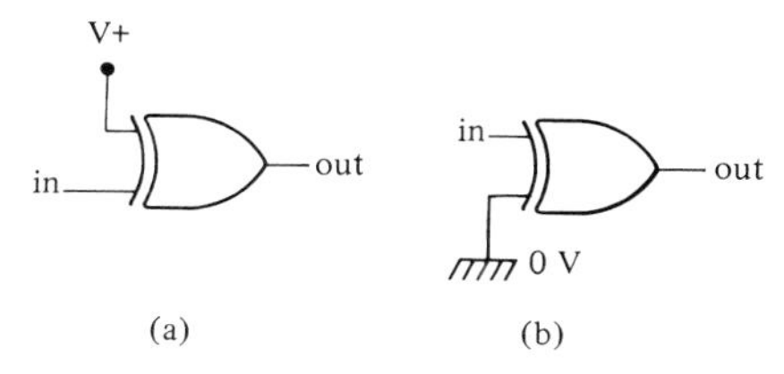

Figure 3.23 *Two-input EX-OR gate connected as (a) inverting and (b) non-inverting amplifier*

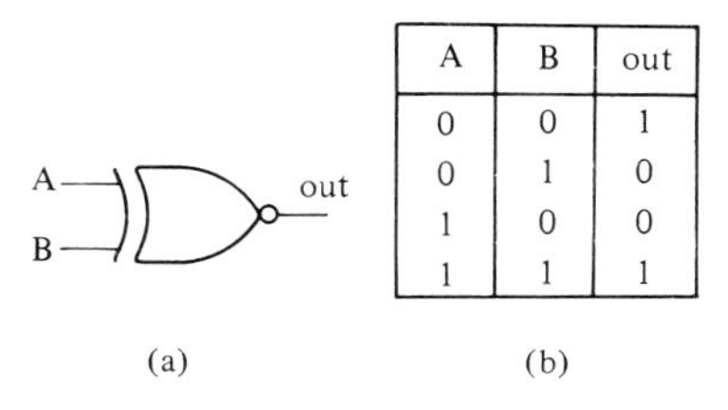

A	B	out
0	0	1
0	1	0
1	0	0
1	1	1

Figure 3.24 *(a) Symbol and (b) truth table of a two-input EX-NOR gate*

Figure 3.24 shows the symbol and truth table of a two-input EX-NOR gate. This logic element is equal to an EX-OR gate with an inverted output. It gives a high output only when both inputs are identical, and is very useful in logic-comparator applications. *Figure 3.25* shows details of the two best known CMOS EX devices, the 4070B quad EX-OR gate and the 4077B quad EX-NOR gate.

Schmitt inverters and gates

CMOS inverters and gates are generally intended to be driven by logic signals that are in either the fully high (logic-1) or fully low (logic-0) states. If inputs

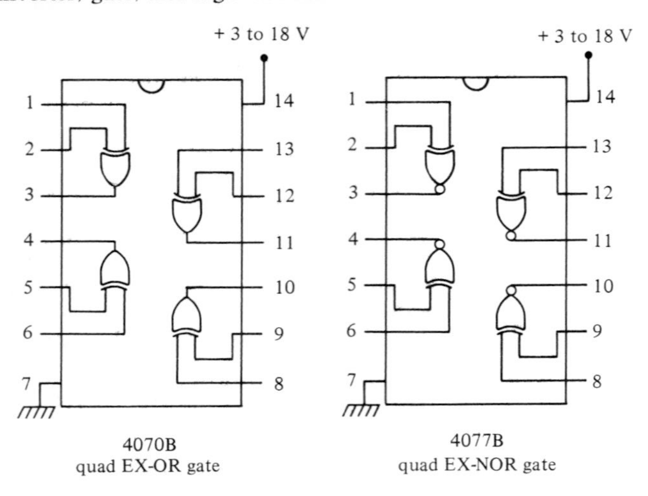

Figure 3.25 *Two popular CMOS EX ICs*

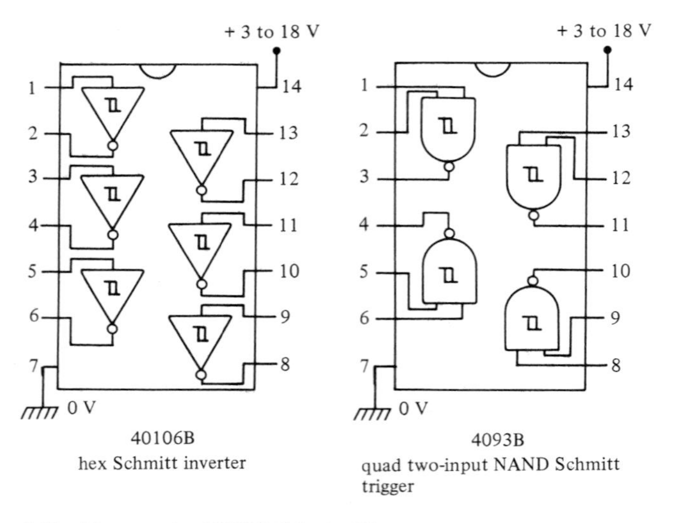

Figure 3.26 *Two popular CMOS Schmitt ICs*

are allowed to linger between these two states for more than a few microseconds, there is a danger that the inverter/gate will become unstable and act as a high-frequency oscillator, thereby generating false output signals.

Consequently, if 'slow' signals are present at one or more of the inputs of a CMOS logic system, these signals must be 'conditioned' (given fast rise and fall times) before they are applied to the actual logic circuitry. The most useful conditioning element is the Schmitt trigger, and *Figure 3.26* gives details of two popular CMOS Schmitt ICs, the 40106B hex Schmitt inverter and the 4093B quad two-input NAND Schmitt trigger.

A programmable gate

Most CMOS logic ICs are dedicated devices; for example, the 4082B is a dual four-input AND gate and can be used as nothing but an AND gate. One very useful exception to this is the 4048B multifunction 'programmable' eight-input gate, which has the functional diagram and outline shown in *Figure 3.27*. This IC has two groups of four input pins, plus an 'expansion' input pin, and is provided with four control (K) pins which enable the user to select the mode of logic operation.

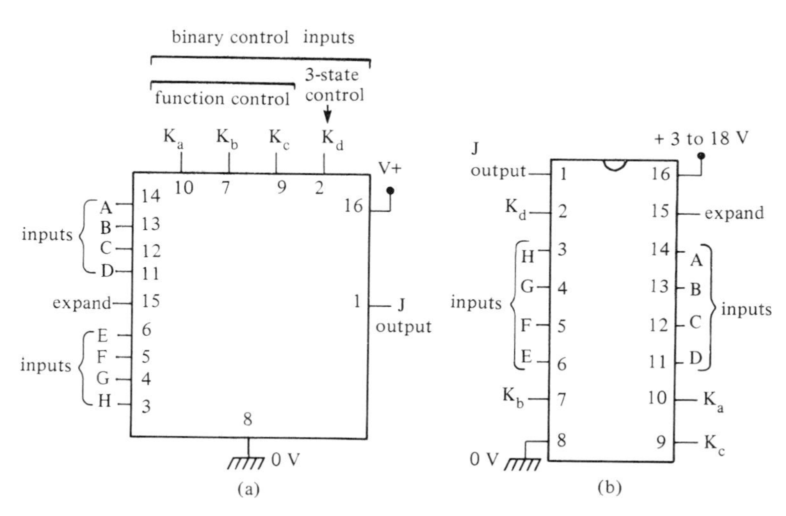

Figure 3.27 *(a) Functional diagram and (b) outline of the 4048B multifunction expandable eight-input gate*

Control input pin K_d (pin 2) enables the user to select either normal (pin 2 high) or high-impedance tristate (pin 2 low) output operation. The remaining three binary control inputs (K_a, K_b, and K_c) enable one of eight different logic functions to be selected, as shown by the table of *Figure 3.28(a)*, which also shows how to connect unwanted inputs in each mode of operation. Thus, to make the 4048B act as a normal six-input OR gate, connect the two unwanted inputs to ground (logic-0), and connect control pins K_a and K_b to ground and

output function	K_a	K_b	K_c	unused inputs
NOR	0	0	0	0
OR	0	0	1	0
OR/AND	0	1	0	0
OR/NAND	0	1	1	0
AND	1	0	0	1
NAND	1	0	1	1
AND/NOR	1	1	0	1
AND/OR	1	1	1	1

K_d = 1 for normal action
= 0 for high-Z outputs
expand input = 0

(a)

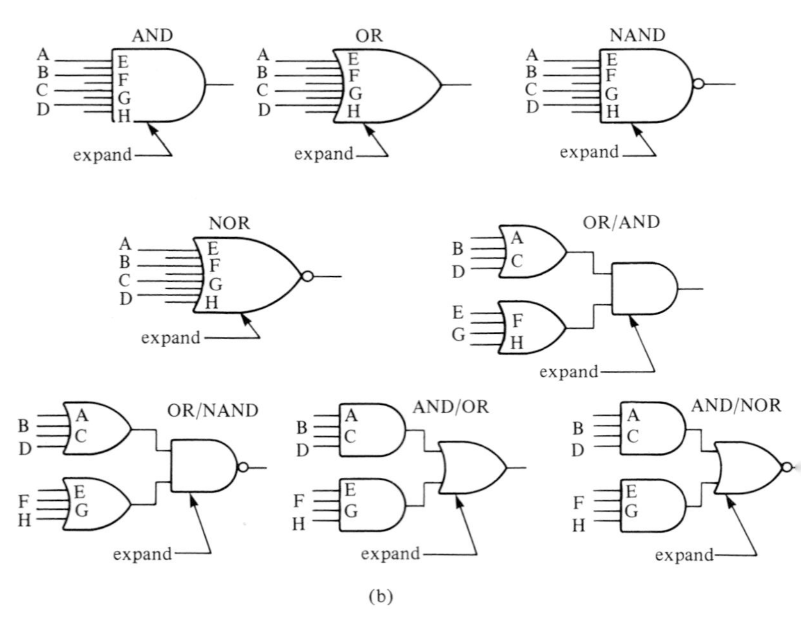

Figure 3.28 *(a) Function table and (b) the eight basic logic configurations of the 4048B multifunction expandable eight-input gate*

K_c and K_d to the positive supply rail. The expand input (pin 15) is normally tied to ground.

Eight different logic functions are available from the 4048B, as shown in

Figure 3.28(b). Note that operation in the AND, OR, NAND, and NOR modes is quite conventional, but that operation in the remaining four modes (OR/AND, OR/NAND, AND/OR, and AND/NOR) is less self-evident. In the latter cases the inputs are broken into two groups of four; each group provides the first part of the logic function, but the *pair* of groups provides the second part of the logic function. Thus, in the OR/AND mode, the circuit gives a high output only if at least one input is present in the A to D group at the same time as at least one input is present in the E to H group.

The expand input terminal of the 4048B enables ICs to be cascaded; thus, for example, two ICs can be made to act as a sixteen-input gate by feeding the output of one IC into the expand terminal of the other. Note when using expanded logic that the input logic feeding the expand terminal is not necessarily the same as the overall logic that is required. Thus an OR expand input is needed for expanded NOR or OR operation, a NAND expand for AND and NAND operation, a NOR expand for OR/AND operation, and an AND expand for AND/OR or AND/NOR operation.

Majority logic

To conclude this chapter, let's take a brief look at a little known logic system known as majority logic, in which the logic unit has an odd number of inputs (three, five, seven etc.) and gives an output only when the majority of inputs

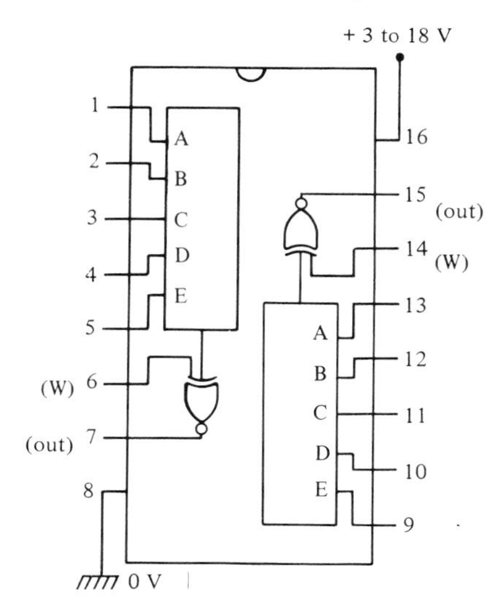

Figure 3.29 *Details of the 4530B dual five-bit majority logic gate*

(two, three, four etc.) are high, irrespective of *which* inputs are active. This type of logic is useful in some special applications, such as in voting machines and semi-intelligent alarms and robotic devices. For example, an alarm bell may sound only if at least two of three detectors indicate a 'fault' condition, or a robot may move only if there is more stimulus to move than there is to stand still.

The best known CMOS majority logic IC is the 4530B dual five-bit unit (*Figure 3.29*), each half of which contains a five-input majority logic element with its output feeding to one input of an EX-NOR gate that has its other input (W) externally available, enabling it to be wired as either an inverting or a non-inverting stage. Thus, when W is tied to logic-1, the EX-NOR stage gives non-inverting action and the output of the element goes high only when the majority of inputs are high. When W is tied to logic-0, the EX-NOR stage gives an inverting action and the output of the element goes high when the majority of inputs are low.

The effective number of inputs of a 4530B can be reduced by wiring half of the unwanted inputs to logic-1 and the other half to logic-0 (*Figure 3.30(a)*). The effective number of inputs can be increased by cascading elements, as shown in *Figure 3.30(b)*, taking the output of one element to one of the inputs of the following element.

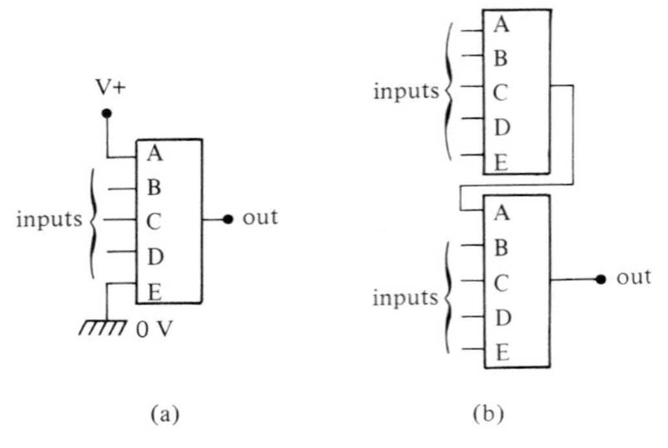

Figure 3.30 *The number of effective inputs of a majority logic circuit can easily be (a) decreased or (b) increased*

The 4530B is actually fairly hard to find. Fortunately, however, majority logic can easily be created by using a 3140 CMOS op-amp in the configuration shown in *Figure 3.31*, which shows a five-input circuit. Here, the op-amp functions as a voltage comparator, with potential divider R_6-R_7 applying half the supply volts to pin 2 of the op-amp, and the five input resistors (which are each connected to either ground or the positive supply rail) form a potential divider that applies a fraction of the supply voltage to pin 3.

Suppose that two input resistors are connected to logic-0 and three resistors

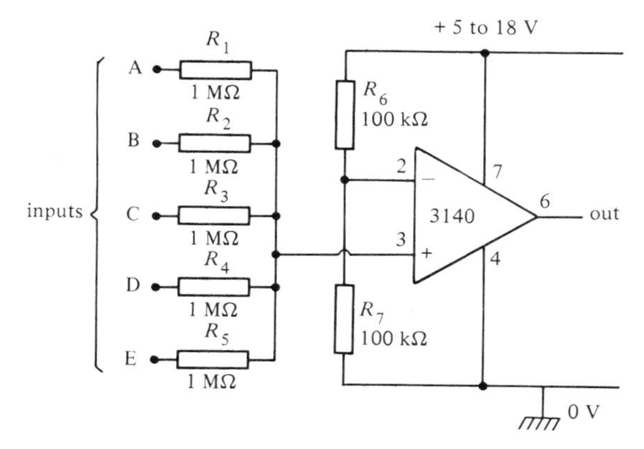

Figure 3.31 *Simple five-input op-amp majority logic gate*

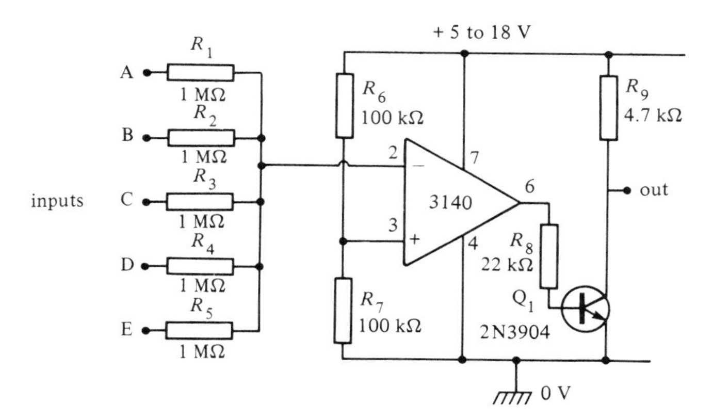

Figure 3.32 *Compound five-input op-amp majority logic gate*

go to logic-1. The three logic-1 resistors have a combined (paralleled) impedance of 333 kΩ, and the two logic-0 resistors have a combined impedance of 500 kΩ, so the resulting potential divider voltage on pin 3 is greater than half the supply volts, causing the output of the op-amp comparator to switch high. If, on the other hand, only two of the five inputs are taken to logic-1, the resulting pin 3 voltage is below half the supply value and the op-amp output is switched low. The circuit thus gives majority logic action.

When 5 per cent resistors are used, the *Figure 3.31* circuit can be given any number of inputs up to a maximum of eleven by simply adding one more 1 MΩ resistor for each new input. The output of the circuit switches fully to zero volts when the output is low, but only rises to within a couple of volts of the supply rail value when the output is high. In most applications this defect is of little importance; it does, however, mean that elements cannot be cascaded to increase the effective total number of inputs. This defect can be overcome by using the alternative compound configuration of *Figure 3.32* in which the output is inverted and level shifted by Q_1, and the inputs to the op-amp are transposed. The output of this circuit switches to within 50 mV of either supply rail, enabling units to be cascaded without limit.

4 Bilateral switches and selectors

A CMOS bilateral switch or transmission gate can be regarded as a near-perfect single-pole single-throw (SPST) electronic switch that can pass analogue or digital signals in either direction and can be turned on (closed) or off (opened) by applying a logic-1 or logic-0 signal to a single high-impedance control terminal. Practical versions of such switches have a near-infinite off impedance, and a typical on impedance in the range 90 to 300 Ω.

Standard CMOS bilateral switches can be switched at rates ranging from near-zero to several megahertz, and have many practical uses. They can, for example, be used to replace mechanical switches in signal-carrying applications; the bilateral switch is DC controlled and placed directly on the PCB, where it is needed, thus eliminating the problems of signal radiation and interaction that normally occur when such signals are mechanically switched via lengthy cables.

At higher frequencies, CMOS bilateral switches can be used in such diverse applications as signal gating, multiplexing, A–D and D–A conversion, digital control of frequency, impedance and signal gain, the synthesis of multigang potentiometers and capacitors, and the implementation of sample-and-hold circuits. Practical examples of most of these applications are shown later in this chapter.

Several types of CMOS multiple bilateral switch IC are available. These range from simple types housing four independently accessible SPST bilateral switches, to fairly complex types housing an array of bilateral switches and logic networks arranged in the form of two independently accessible single-pole eight-way bilateral switches or multiplexers/demultiplexers. Before we take a detailed look at the range of such ICs, let's look at the basis operating principles and terminology of the bilateral switch.

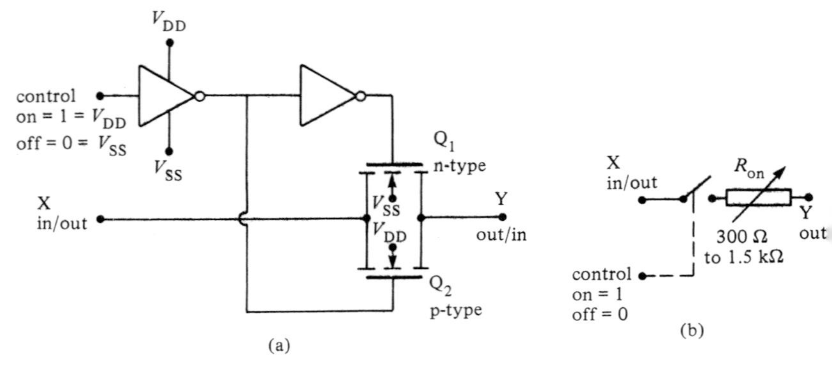

Figure 4.1 *(a) Basic circuit and (b) equivalent circuit of the simple CMOS bilateral switch*

Basic principles

Figure 4.1 shows (a) the basic circuit and (b) the equivalent circuit of a simple CMOS bilateral switch. Here, an n-type and a p-type MOSFET are effectively wired in inverse parallel (drain-to-source and source-to-drain), but have their gates biased in anti-phase from the control terminal via a pair of inverters. When the control signal is at the logic-0 level, the gate of Q_2 is driven to V_{DD} and the gate of Q_1 is driven to V_{SS}; under this condition both MOSFETs are cut off, and an effective open circuit exists between the X and Y points of the circuit. When, on the other hand, the control signal is set at the logic-1 level, the gate of Q_2 is driven to V_{SS} and the gate of Q_1 is driven to V_{DD}, and under this condition both MOSFETs are driven to saturation, and a near short-circuit exists between the X and Y points.

Note that, when Q_1 and Q_2 are saturated, signal currents can flow in either direction between the X and Y terminals, provided that the signal voltages are within the V_{SS}-to-V_{DD}limits. Each of the X and Y terminals can thus be used as either an in or an out terminal.

In practice, Q_1 and Q_2 exhibit a finite resistance (R_{on}) when they are saturated, and in this simple circuit the actual value of R_{on} may vary from 300 Ω to 1.5 kΩ, depending on the magnitude of the V_{SS}-to-V_{DD} supply voltage and on the magnitude and polarity of the actual input signal. The simple bilateral switch can thus be represented by the equivalent circuit of *Figure 4.1(b)*.

Figure 4.2 shows an improved version of the CMOS bilateral switch, together with its equivalent circuit. This circuit is similar to the above, except for the addition of a second bilateral switch (Q_3-Q_4) that is wired in series with Q_5, with the 'well' of Q_1 tied to the Q_5 drain. These modifications cause Q_1's

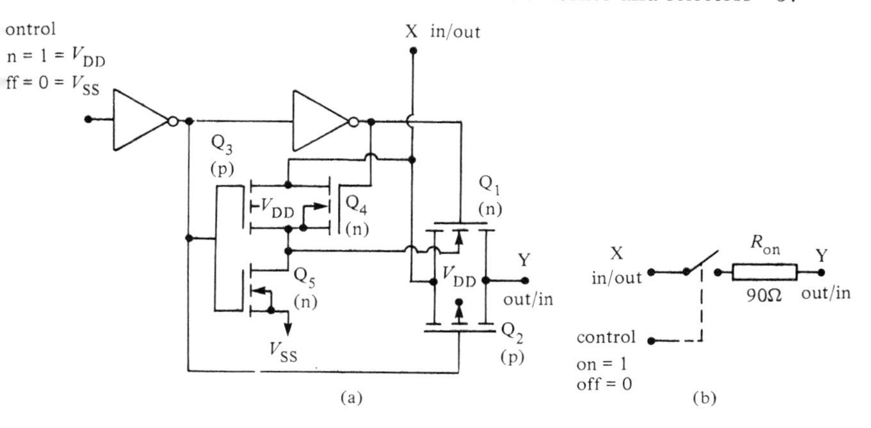

Figure 4.2 *(a) Basic circuit and (b) equivalent circuit of the improved CMOS bilateral switch*

well to switch to V_{SS} when the Q_1-Q_2 bilateral switch is off, but to be tied to the X input terminal when the switch is on. This modification reduces the on resistance of the Q_1-Q_2 bilateral switch to about 90 Ω and virtually eliminates variations in its value with varying signal voltages etc., as indicated by the equivalent circuit of *Figure 4.2(b)*. The only disadvantage of the *Figure 4.2* circuit is that it has a slightly lower leakage resistance than that of *Figure 4.1*.

Switch biasing

A CMOS bilateral switch can be used to switch or gate either digital or analogue signals, but must be correctly biased to suit the type of signal being controlled. *Figure 4.3* shows the basic ways of activating and biasing the bilateral switch. *Figure 4.3(a)* shows that the switch can be turned on (closed) by taking the control terminal to V_{DD}, or turned off (open) by taking the control terminal to V_{SS}.

In digital signal switching applications (*Figure 4.3(b)*) the bilateral switch can be used with a single-ended supply, with V_{SS} to zero volts and V_{DD} at a positive value equal to (or greater than) that of the digital signal (up to a maximum of +18 volts). In analogue switching applications (*Figure 4.3(c)*) a split supply (either true or effective) must be used so that the signal is held at a mean value of 'zero' volts; the positive supply rail goes to V_{DD}, which must be greater than the peak positive voltage value of the input signal, and the negative rail goes to V_{SS} and must be greater than the peak negative value of the input signal; the supply values are limited to plus or minus 9 volts

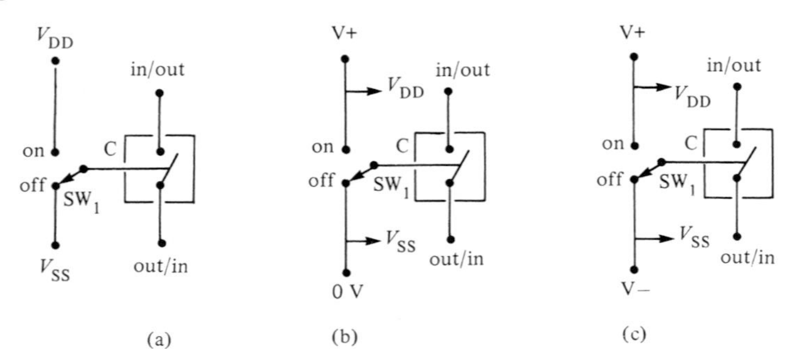

Figure 4.3 *Basic methods of turning the bilateral switch on and off (a), and power connections for use with (b) digital and (c) analogue in/out signals*

maximum. Typically, the bilateral switch introduces less than 0.5 per cent of signal distortion when used in the analogue mode.

Logic-level conversion

Note from the above description of the analogue system that, if a split supply is used, the switch control signal must switch to the positive rail to turn the bilateral switch on, and to the negative rail to turn the switch off. This arrangement is inconvenient in many practical applications. Consequently, some CMOS bilateral switch ICs (notably the 4051B to 4053B family) have built-in logic-level conversion circuitry which enables the bilateral switches to be controlled by a digital signal that switches between zero (V_{SS}) and positive (V_{DD}) volts, while still using split supplies to give correct biasing for analogue operation, as shown in *Figure 4.4*

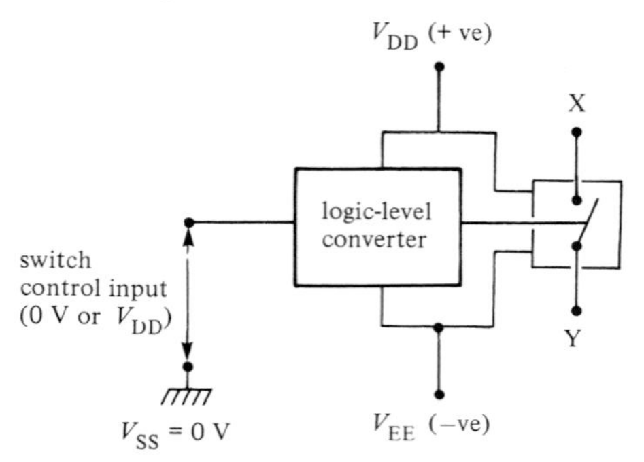

Figure 4.4 *Some ICs feature internal logic-level conversion, enabling an analogue switch to be controlled via a single-ended input*

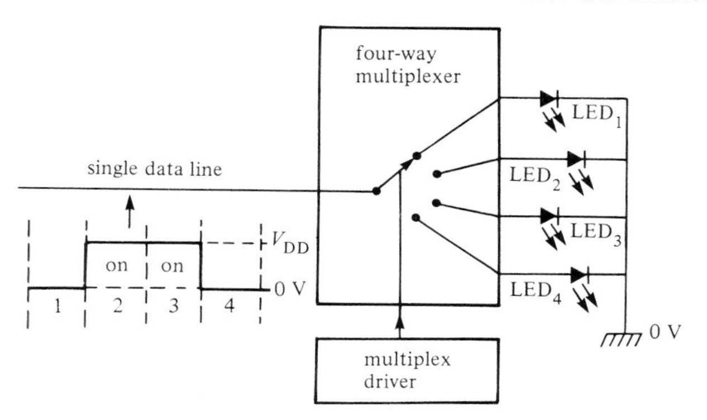

Figure 4.5 *A four-way multiplexer used to control four LEDs via a single data line*

Multiplexing/demultiplexing

A multiplexer can be regarded as any system that enables information from a single data line to be distributed – on a sequential time-share basis – to a number *n* of independent data lines. *Figure 4.5*, for example, shows how a four-way multiplexer (represented by a four-way switch) can be used to control (turn on or off) four LEDs down a single data line.

In *Figure 4.5*, assume that the multiplex driver continuously sequences the multiplexer through the 1-2-3-4 cycle at a fairly rapid rate, and is synchronized to the 1-2-3-4 segments of the data line. Thus, in each cycle, in the 1 period LED_1 is off; in the 2 period LED_2 is on; in the 3 period LED_3 is on, and in the 4 period LED_4 is off. The state of each of the four LEDs is thus controlled via the logic bit of the single (sequentially time-shared) data line.

A demultiplexer is the opposite of a multiplexer. It enables information from a number *n* of independent data lines to be sequentially applied to a single data line. *Figure 4.6* shows how a four-way demultiplexer can be used to feed three independent 'voice' signals down a single cable, and how a multiplexer can be used to convert signals back into three independent voice signals at the other end. In practice, each 'sample' period of the data line must be short relative to the period of the highest voice frequency; period 1 is used to synchronize the signals at the two ends of the data line.

From the above description it can be seen that a CMOS *n*-channel multiplexer can be regarded as a single-pole *n*-way bilateral switch, and that a CMOS multiplexer can be converted into a demultiplexer by simply transposing the notations of the input and output terminals. An *n*-way single-pole bilateral switch can thus be described as an *n*-channel multiplexer/demultiplexer.

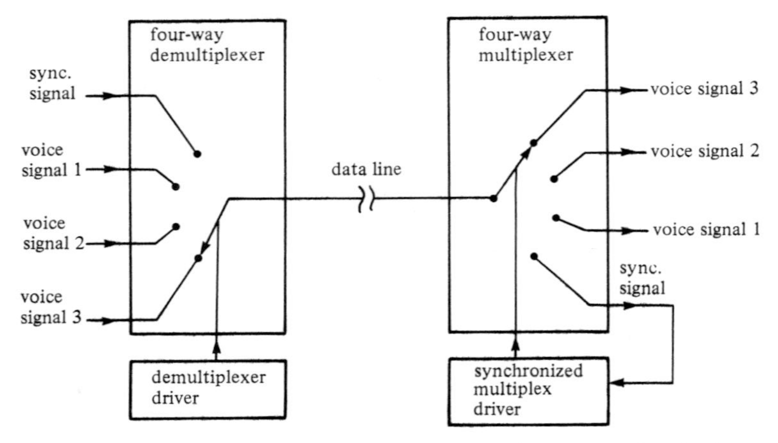

Figure 4.6 *A four-way multiplexer/demultiplexer combination used to feed three independent voice signals through a single data line*

Practical ICs

There are three major families of CMOS bilateral switch ICs. The best known of these comprises the 4016B/4066B types, which are quad bilateral switches, each housing four independently accessible SPST bilateral switches, as shown in *Figure 4.7*. The 4016B uses the simple construction shown in *Figure 4.1*, and is recommended for use in sample-and-hold applications where low leakage impedance is of prime importance. The 4066B uses the improved type of construction of *Figure 4.2*, and is recommended for use in all applications where a low on resistance is of prime importance.

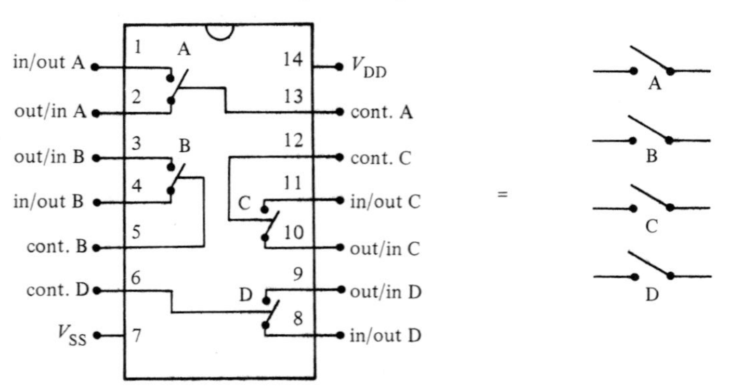

Figure 4.7 *The 4016B and 4066B quad bilateral switches each act as four independent SPST switches*

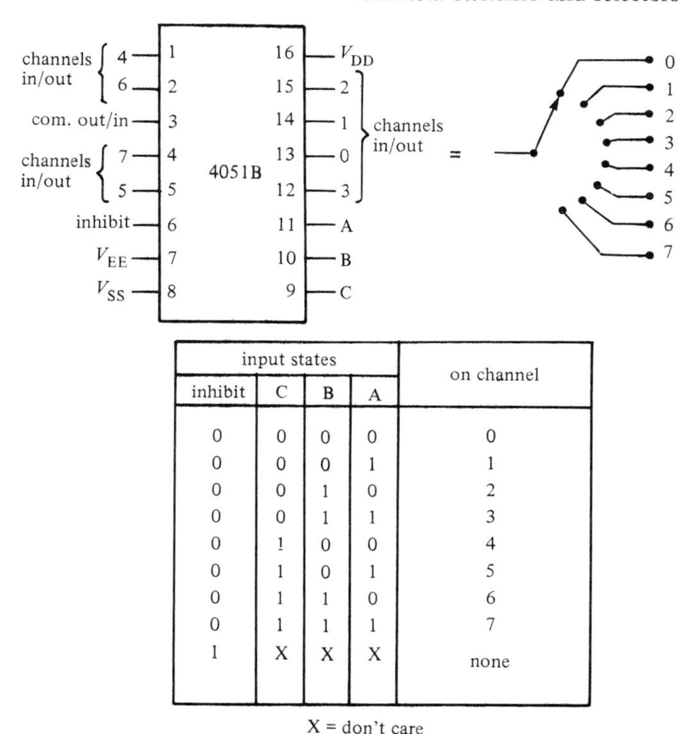

input states				on channel
inhibit	C	B	A	
0	0	0	0	0
0	0	0	1	1
0	0	1	0	2
0	0	1	1	3
0	1	0	0	4
0	1	0	1	5
0	1	1	0	6
0	1	1	1	7
1	X	X	X	none

X = don't care

Figure 4.8 *The 4051B eight-channel multiplexer/demultiplexer acts as a single-pole eight-way switch*

The second family of ICs comprises the 4051B to 4053B types (*Figures 4.8* to *4.10*). These are multichannel multiplexer/demultiplexer ICs featuring built-in logic-level conversion. These ICs have three 'power supply' terminals (V_{DD}, V_{SS}, and V_{EE}). In all applications, V_{DD} is taken to the positive supply rail and V_{SS} is grounded, and all digital control signals (for channel select, inhibit, etc.) use these two terminals as their logic reference values, i.e. logic-1 = V_{DD} and logic-0 = V_{SS}. In digital signal processing applications, terminal V_{EE} is grounded (tied to V_{SS}.) In analogue signal processing applications, V_{EE} must be taken to a negative supply rail; ideally, $V_{EE} = -V_{DD}$. In all cases, the V_{EE}-to-V_{DD} voltage must be limited to 18 volts maximum.

The 4051B (*Figure 4.8*) is an eight-channel multiplexer/demultiplexer, and can be regarded as a single-pole, eight-way bilateral switch. The IC has three binary control inputs (A, B, and C) and an inhibit input. The three binary signals select the one of the eight channels to be turned on, as shown in the table.

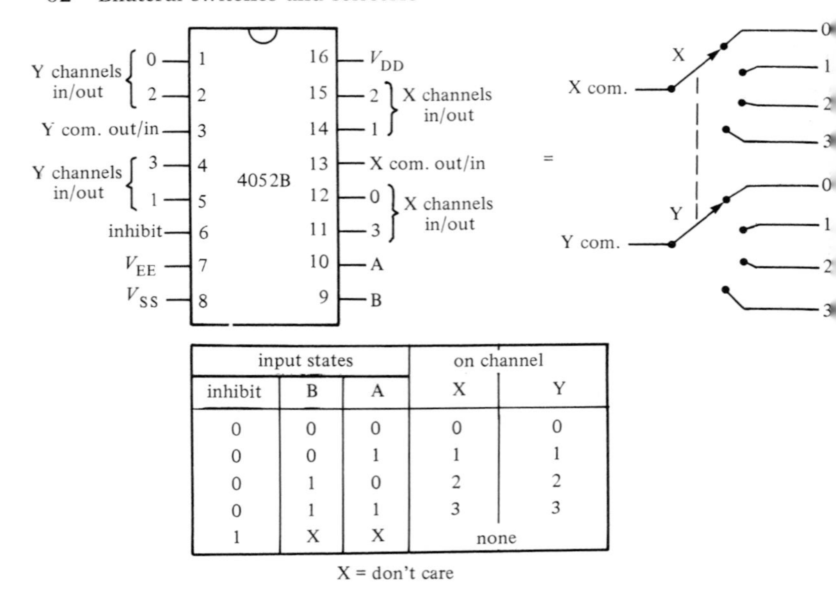

input states			on channel	
inhibit	B	A	X	Y
0	0	0	0	0
0	0	1	1	1
0	1	0	2	2
0	1	1	3	3
1	X	X	none	

X = don't care

Figure 4.9 *The 4052B differential four-channel multiplexer/demultiplexer acts as a ganged two-pole four-way switch*

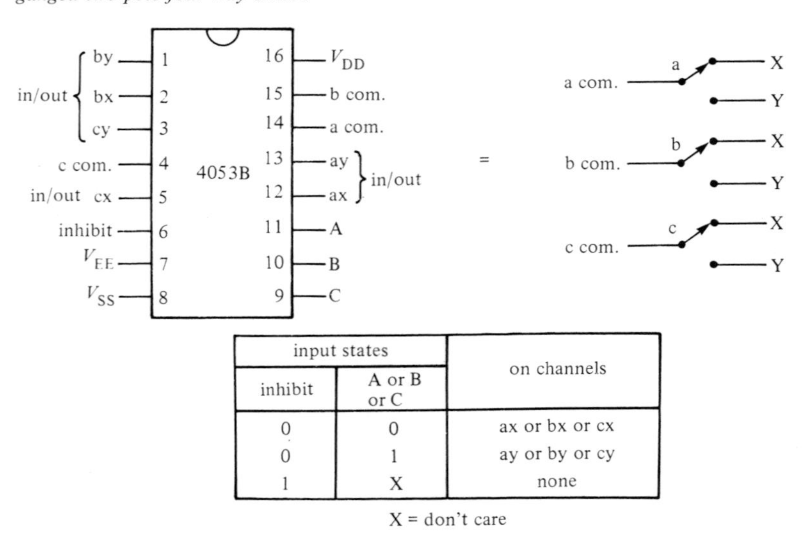

input states		on channels
inhibit	A or B or C	
0	0	ax or bx or cx
0	1	ay or by or cy
1	X	none

X = don't care

Figure 4.10 *The 4053B triple two-channel multiplexer/demultiplexer acts as three independent single-pole two-way switches*

The 4052B (*Figure 4.9*) is a differential four-channel multiplexer/demultiplexer or ganged two-pole four-way bilateral switch, and the 4053B (*Figure 4.10*) is a triple two-channel multiplexer/demultiplexer or set of three independent single-pole three-way bilateral switches.

The final family of devices comprises the 4067B and 4097B multiplexer/demultiplexer types (*Figures 4.11* and *4.12*). These devices can be used in both analogue and digital applications, but do not feature built-in logic-level conversion. The 4067B is a sixteen-channel device, and can be regarded as a single-pole sixteen-way bilateral switch. The 4097B is a differential eight-channel device, and can be regarded as a ganged two-pole eight-way bilateral switch. Each IC is housed in a 24-pin dual-in-line DIL package.

Using 4016B/4066B ICs

The 4016B and 4066B are very versatile ICs, but a few simple precautions must be taken when using them:

1 Input and switching signals must never be allowed to rise above V_{DD} or below V_{SS}.

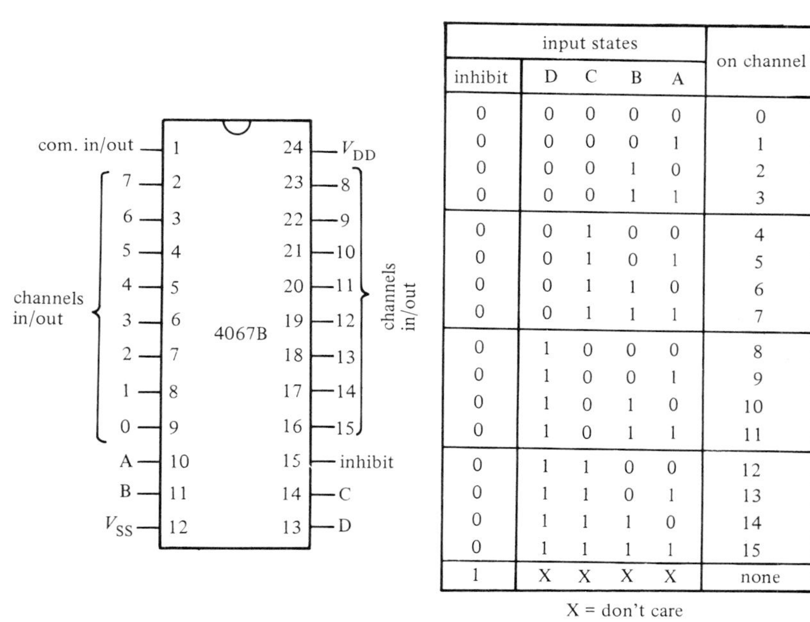

input states					on channel
inhibit	D	C	B	A	
0	0	0	0	0	0
0	0	0	0	1	1
0	0	0	1	0	2
0	0	0	1	1	3
0	0	1	0	0	4
0	0	1	0	1	5
0	0	1	1	0	6
0	0	1	1	1	7
0	1	0	0	0	8
0	1	0	0	1	9
0	1	0	1	0	10
0	1	0	1	1	11
0	1	1	0	0	12
0	1	1	0	1	13
0	1	1	1	0	14
0	1	1	1	1	15
1	X	X	X	X	none

X = don't care

Figure 4.11 *The 4067B sixteen-channel multiplexer/demultiplexer acts as a single-pole sixteen switch*

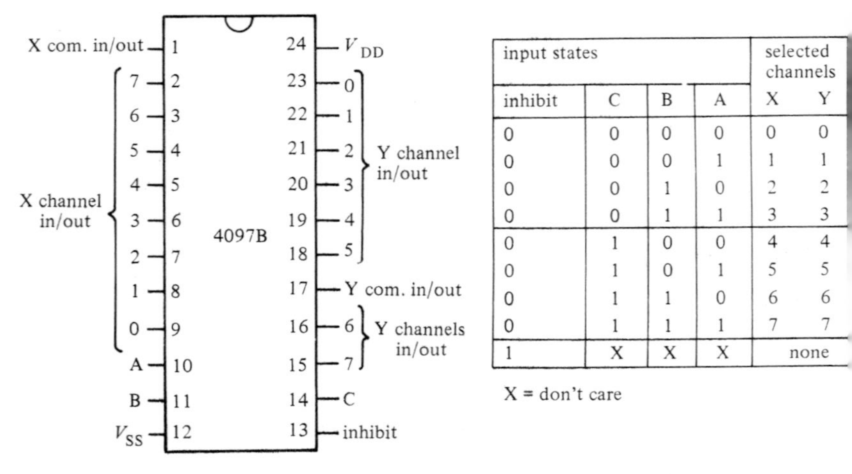

input states				selected channels	
inhibit	C	B	A	X	Y
0	0	0	0	0	0
0	0	0	1	1	1
0	0	1	0	2	2
0	0	1	1	3	3
0	1	0	0	4	4
0	1	0	1	5	5
0	1	1	0	6	6
0	1	1	1	7	7
1	X	X	X	none	

X = don't care

Figure 4.12 *The 4097B differential eight-channel multiplexer/demultiplexer acts as a ganged two-pole eight-way switch*

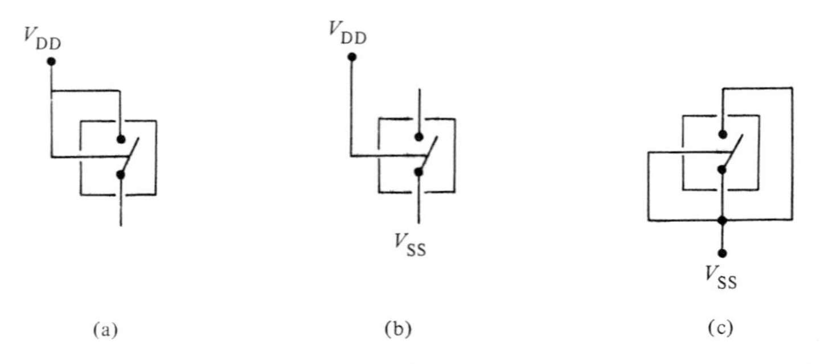

Figure 4.13 *Unused sections of the 4066B must be disabled, using any one of the connections shown here*

2 Each unused section of the IC must be disabled (see *Figure 4.13*) either by taking its control terminal to V_{DD} and wiring one of its switch terminals to V_{DD} or V_{SS}, or by taking all three terminals to V_{SS}.

Figures 4.14 to *4.19* show some simple applications of the 4066B (or 4016B). *Figure 4.14* shows the device used to implement the four basic switching functions of SPST, SPDT, DPST, and DPDT. *Figure 4.14(a)* shows the SPST connection, which we have already discussed. The SPDT function (*Figure 4.14(b)*) is implemented by wiring an inverter stage (a 4001B or 4011B etc.) between the ICI_a and ICI_b control terminals. The DPST switch (*Figure 4.14(c)*) is simply two SPST switches sharing a common control terminal, and

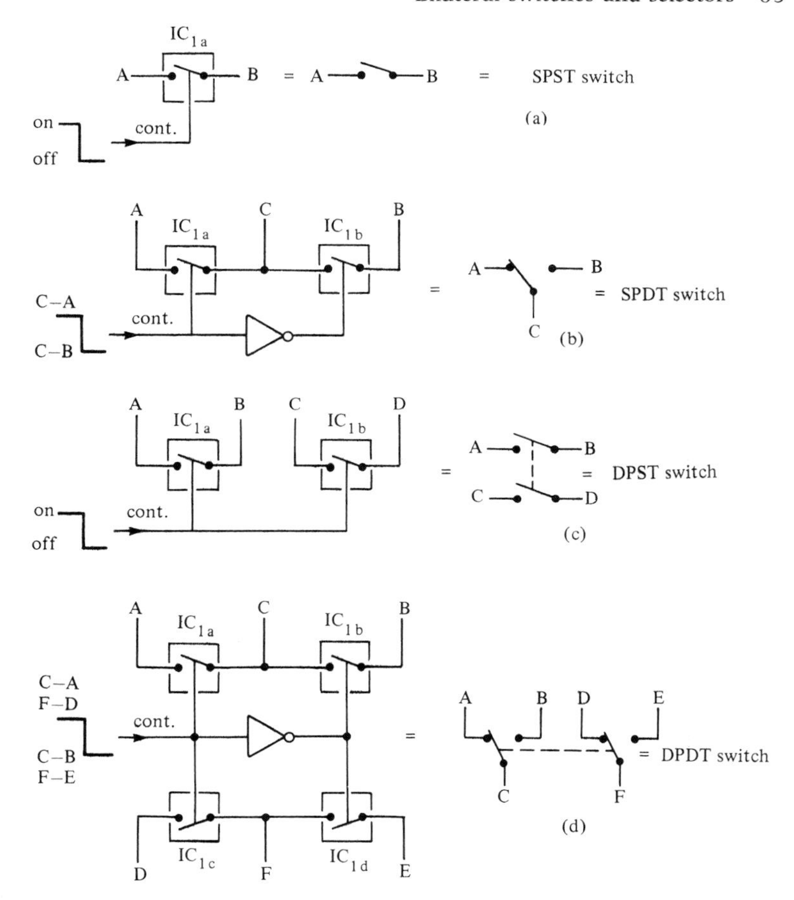

Figure 4.14 *Implementation of the four basic switching functions via the 4066B (IC_1)*

the DPDT switch (*Figure 4.14(d)*) is two SPDT switches sharing an inverter stage in the control line.

Note that the switching functions of *Figure 4.14* can be expanded or combined in any desired way by using more IC stages. Thus, a ten-pole two-way switch can be made by using five of the *Figure 4.14(d)* circuits with their control lines tied together.

Each 4066B bilateral switch has a typical on resistance of about 90 Ω. *Figure 4.15* shows how four standard switch elements can be wired in parallel to make a single switch with a typical on resistance of only 22.5 Ω.

Figures 4.16 to *4.19* shows ways of using a bilateral switch as a self-latching

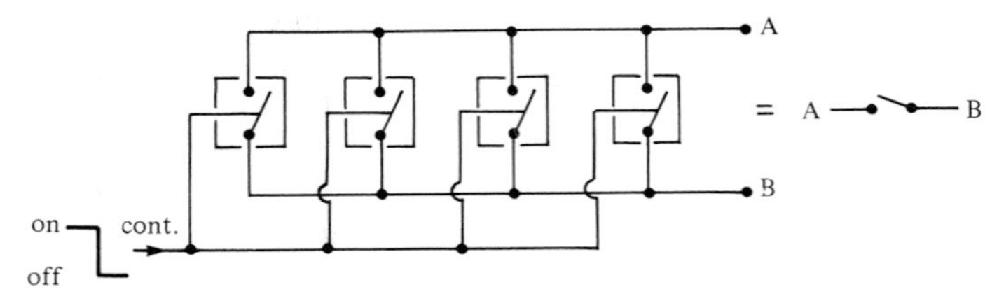

Figure 4.15 *This SPST switch has a typical on resistance of only 22.5 ohms*

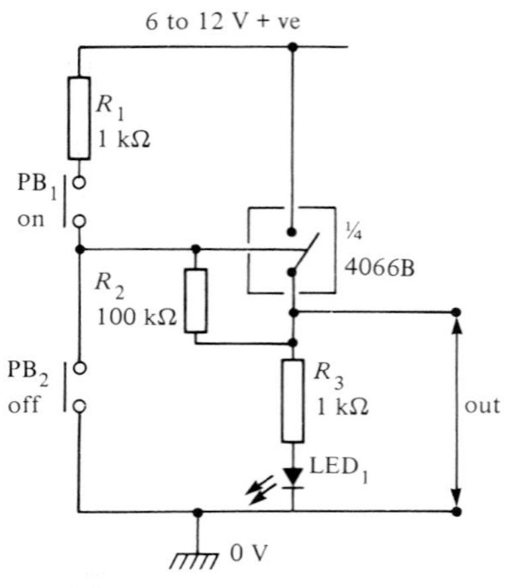

Figure 4.16 *Latching push-button switch*

device. In these circuits the switch current flows to ground via R_3, and the control terminal is tied to the top of R_3 via R_2. Thus, in *Figure 4.16*, when PB_1 is briefly closed the control terminal is pulled to the positive rail and the bilateral switch closes. With the bilateral switch closed, the top of R_3 is at supply line potential and, since the control terminal is tied to R_3 via R_2, the bilateral switch is thus latched on. Once latched, the switch can only be turned off again by briefly closing PB_2, at which point the bilateral switch opens and the R_3 voltage falls to zero. Note here that LED_1 merely indicates the state of the bilateral switch, and R_1 prevents supply line short-circuits if PB_1 and PB_2 are both closed at the same time.

Figure 4.17 shows how the above circuit can be made to operate as a latching touch-operated switch by increasing R_2 to 10 MΩ and using R_4-C_1 as a 'hum' filter.

Figures 4.18 and *4.19* show alternative ways of using the *Figure 4.16* circuit to connect power to external circuitry. The *Figure 4.18* circuit connects the power via a voltage-follower stage, and the *Figure 4.19* design connects the power via a common-emitter amplifier.

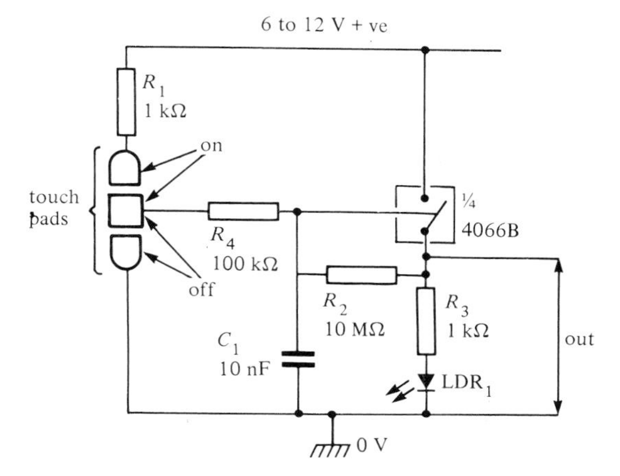

Figure 4.17 *Latching touch switch*

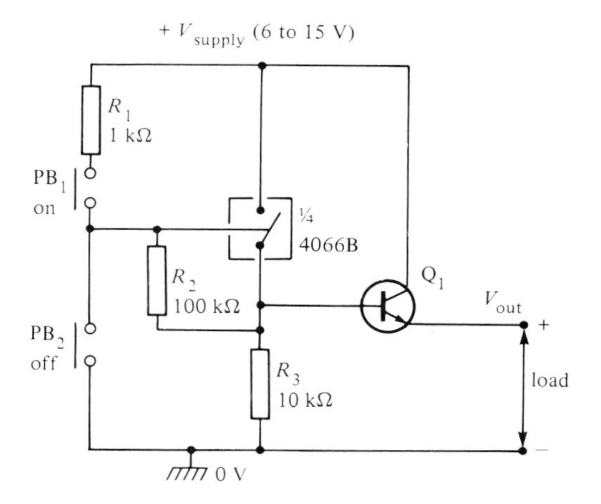

Figure 4.18 *Latching push-button power switch*

Digital control circuits

Bilateral switches can be used to digitally control or vary effective values of resistance, capacitance, impedance, amplifier gain, oscillator frequency, etc. in any desired number of discrete steps. *Figure 4.20* shows how the four switches of a single 4066B can, by either short circuiting or not short circuiting individual resistors in a chain, be used to vary the effective total value of the resistance chain in sixteen digitally controlled steps of 10 kΩ each.

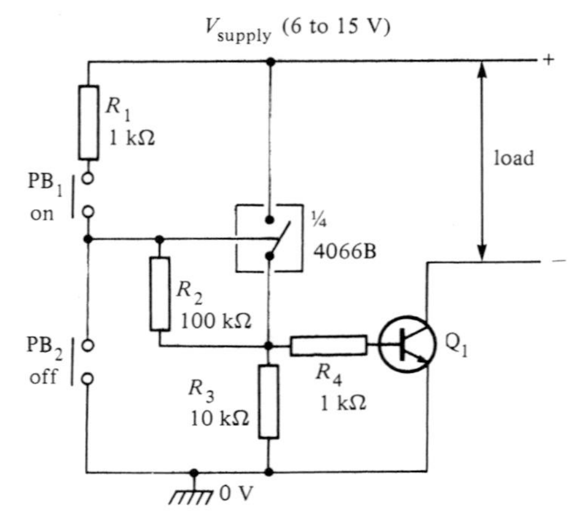

Figure 4.19 *Alternative version of the latching power switch*

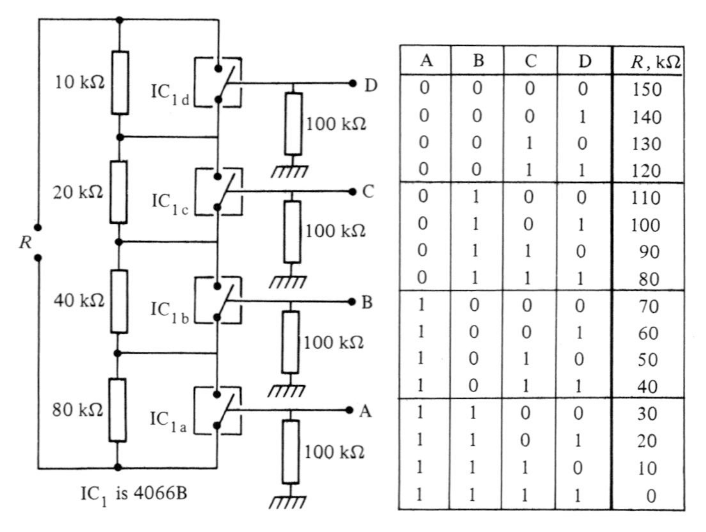

A	B	C	D	R, kΩ
0	0	0	0	150
0	0	0	1	140
0	0	1	0	130
0	0	1	1	120
0	1	0	0	110
0	1	0	1	100
0	1	1	0	90
0	1	1	1	80
1	0	0	0	70
1	0	0	1	60
1	0	1	0	50
1	0	1	1	40
1	1	0	0	30
1	1	0	1	20
1	1	1	0	10
1	1	1	1	0

Figure 4.20 *Sixteen-step digital control of resistance.* R *can be varied from zero to 150 kΩ in 10 kΩ steps*

In practice, the step magnitudes of the *Figure 4.20* circuit can be given any desired value (determined by the value of the smallest resistor) as long as the four resistors are kept in the ratio 1:2:4:8. The number of steps can be increased by adding more resistor/switch stages; thus, a six-stage circuit (with resistors in the ratio (1:2:4:8:16:32) will give resistance variation in 64 steps.

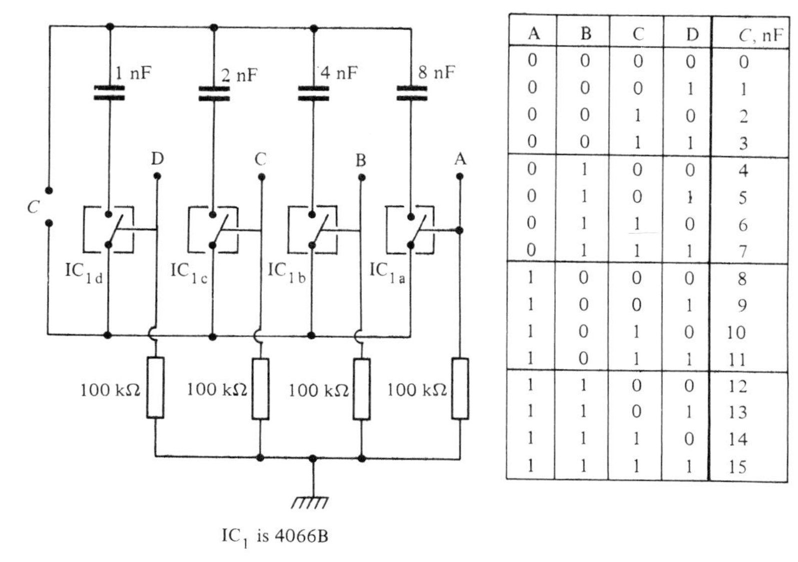

A	B	C	D	C, nF
0	0	0	0	0
0	0	0	1	1
0	0	1	0	2
0	0	1	1	3
0	1	0	0	4
0	1	0	1	5
0	1	1	0	6
0	1	1	1	7
1	0	0	0	8
1	0	0	1	9
1	0	1	0	10
1	0	1	1	11
1	1	0	0	12
1	1	0	1	13
1	1	1	0	14
1	1	1	1	15

Figure 4.21 *Sixteen-step digital control of capacitance.* C *can be varied from zero to 15 nF in 1nF steps*

Figure 4.21 shows how four switches can be used to make a digitally controlled capacitor that can be varied in sixteen steps of 1 nF each. Again, the circuit can be expanded to give more steps by simply adding more stages.

Note that in the *Figures 4.20* and *4.21* circuits the resistor/capacitor values can be controlled by operating the 4066B switches manually, or automatically via simple logic networks, or via up/down counters (see Chapter 8), or via microprocessor control, etc.

The circuits of *Figures 4.20* and *4.21* can be combined in a variety of ways to make digitally controlled impedance and filter networks etc. *Figure 4.22*, for

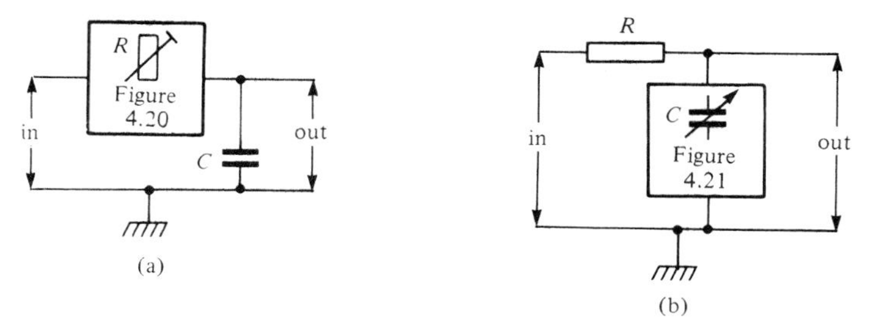

Figure 4.22 *Alternating ways of using* Figure 4.20 *or* Figure 4.21 *to make a digitally controlled first-order low-pass filter*

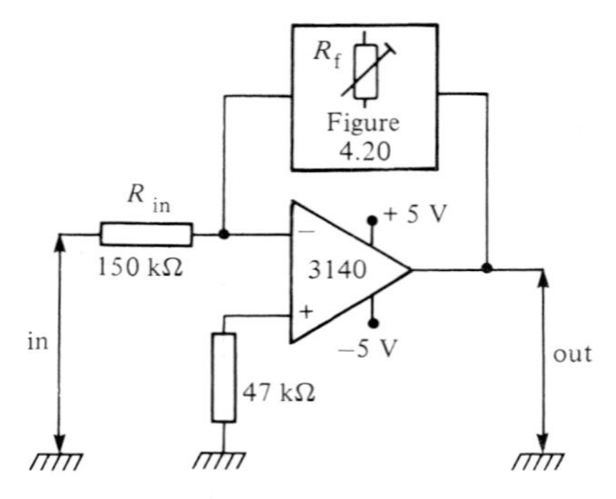

Figure 4.23 *Digital control of gain, using the* Figure 4.20 *circuit. Gain is variable from zero to unity in sixteen steps*

example, shows two alternative ways of using them to make a digitally controlled first-order low-pass filter.

Digital control of amplifier gain can be obtained by hooking the *Figure 4.20* circuit into the feedback or input path of a standard op-amp inverter circuit, as shown in *Figures 4.23* and *4.24*. The gain of such a circuit equals R_f/R_{in}, where R_f is the feedback resistance and R_{in} is the input resistance. Thus in *Figure 4.23* the gain can be varied from zero to unity in sixteen steps of 1/15th each, giving a sequence of 0/15 (i.e. zero), 1/15, 2/15, etc. up to 14/15 and (finally) 15/15 (i.e. unity).

In the *Figure 4.24* circuit the gain can be varied from unity to 16 in sixteen steps, giving a gain sequence of 1, 2, 3, 4, 5, etc. Note that in both of these circuits the op-amp uses a split power supply, so the 4066B control voltage must switch between the negative and positive supply rails.

Figure 4.25 shows how the *Figure 4.20* circuit can be used to vary the frequency of a 555 astable oscillator in sixteen discrete steps. Finally, *Figure 4.26* shows how three bilateral switches can be used to implement

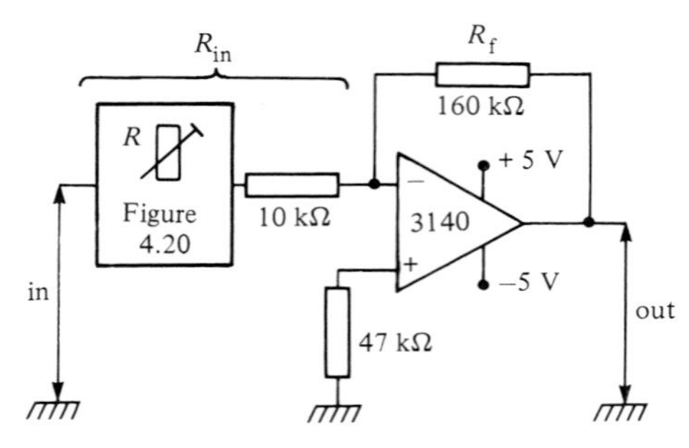

Figure 4.24 *Digital control of gain, using the* Figure 4.20 *circuit. Gain is variable from unity to 16 in sixteen steps*

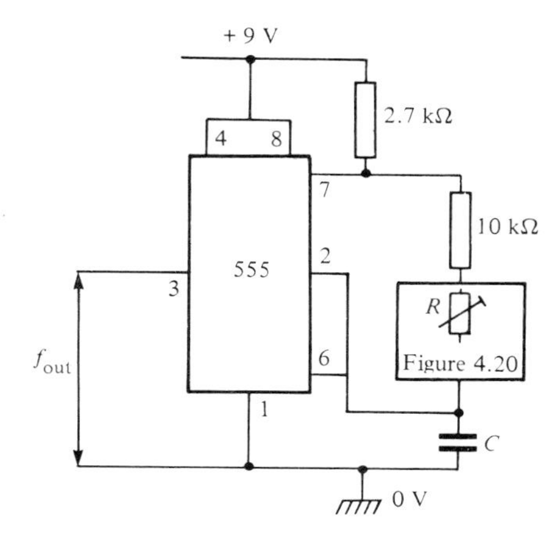

Figure 4.25 *Digital control of 555 astable frequency, in sixteen steps*

digital control of decade range selection of a 555 astable oscillator. Here, only one of the switches must be turned on at a time. Naturally, the circuits of *Figures 4.25* and *4.26* can easily be combined, to form a wide-range oscillator that can be digitally controlled via a microprocessor or other device.

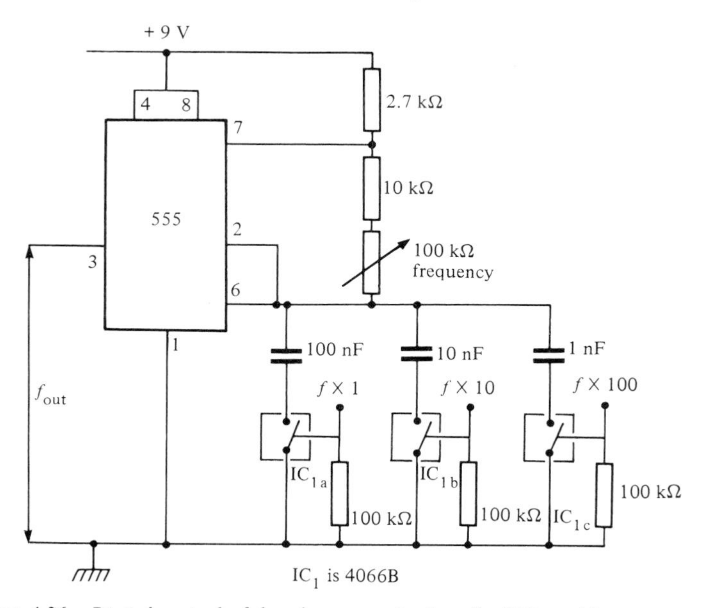

Figure 4.26 *Digital control of decade range selection of a 555 astable*

Synthesized multigang potentiometers

One of the most useful applications of the bilateral switch is in synthesizing multigang rheostats, potentiometers, and variable capacitors in AC signal processing circuitry. The synthesizing principle is quite simple and is illustrated in *Figure 4.27*, which shows the circuit of a four-gang 10–100 kΩ rheostat for use at signal frequencies up to about 15 kHz.

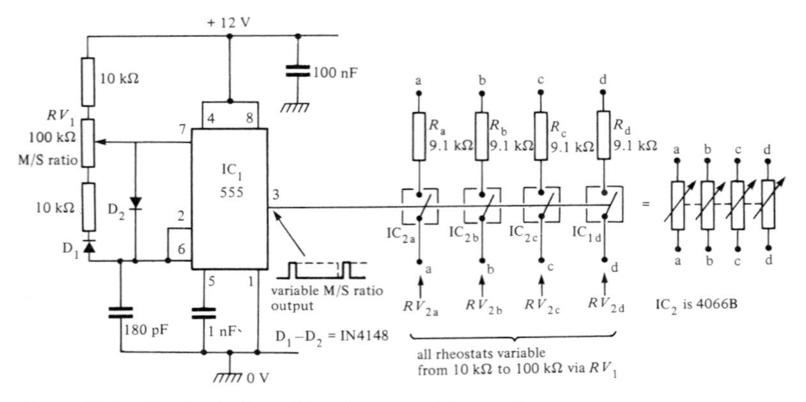

Figure 4.27 *Synthesized precision four-gang 'rheostat'*

Here, the 555 is used to generate a 50 kHz rectangular wave form that has its mark/space (M/S) ratio variable from 11:1 to 1:11 via RV_1, and this waveform is used to control the switching of the 4066B stages. All of the 4066B switches are fed with the same control waveform, and each switch is wired in series with a range resitor (R_a, R_b, etc), to form one gang of the 'rheostat' between the aa, bb, cc, and dd terminals.

Remembering that the switching rate of this circuit is fast (50 kHz) relative to the intended maximum signal frequency (15 kHz), it can be seen that the *mean* or effective value (when integrated over a few switching cycles) of each rheostat resistance can be varied via M/S ratio control RV_1.

Thus, if IC_{2a} is closed for 90 per cent and open for 10 per cent of each duty cycle (M/S ratio = 9:1), the apparent (mean) value of the aa resistance will be 10 per cent greater than R_a, i.e. 10 kΩ. If the duty cycle is reduced to 50 per cent, the apparent R_a value will double to 18.2 kΩ. If the duty cycle is further decreased, so that IC_{2a} is closed for only 10 per cent of each duty cycle (M/S ratio = 1:9), the apparent value of R_a will increase by a decade to 91 kΩ. Thus the apparent value of each gang of the rheostat can be varied via RV_1.

There are some important points to note about the above circuit. First, it can be given any desired number of 'gangs' by simple adding an appropriate number of switch stages and range resistors. Since all switches are controlled by the same M/S ratio waveform, perfect tracking is automatically assured between the gangs. Individual gangs can be given different ranges, without affecting the tracking, by giving them different range resistor values. Also, note that the 'sweep' range and 'law' of the rheostats can be changed by simply altering the characteristics of the M/S ratio generator.

Note that in the above circuit the switching control frequency *must* be far higher than the maximum signal frequency that is to be handled, or the circuit will not perform correctly.

The rheostat circuit of *Figure 4.27* can be made to function as a multigang variable capacitor by using ranging capacitors in place of ranging resistors. In this case, however, the apparent capacitance value decreases as the duty cycle is decreased.

The *Figure 4.27* principle can be expanded to make synthesized multigang potentiometers by using the basic technique shown in *Figure 4.28*. Here, in each gang, two rheostats are wired in series but have their switch control signals fed in anti-phase, so that one rheostat value increases as the other decreases, thus giving a variable potential divider action. This basic circuit can be expanded to incorporate any desired number of gangs by simply adding more double-rheostat stages.

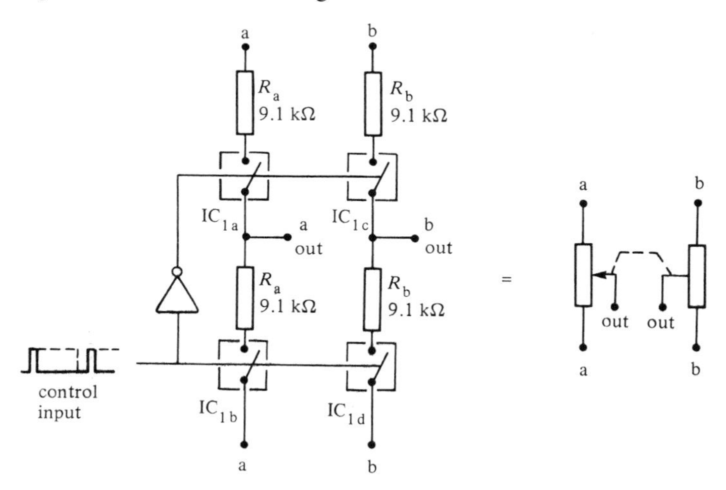

Figure 4.28 *Synthesized precision two-gang 'potentiometer'*

Miscellaneous applications

To complete this look at the CMOS bilateral switch, *Figures 4.29* and *4.30* show a couple of miscellaneous fast-switching applications of the device. *Figure 4.29* shows how it can be used as a sample-and-hold element. The 3140

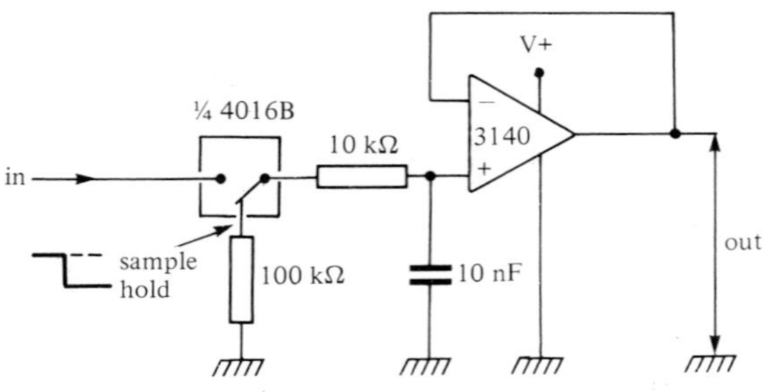

Figure 4.29 *Using a bilateral switch as a sample-and-hold element*

CMOS op-amp is used as a voltage follower and has a near-infinite input impedance, and the 4016B switch also has a near-infinite impedance when open. Thus, when the 4016B is closed, the 10 nF capacitor rapidly follows all variations in input voltage, but when the switch opens the prevailing capacitance charge is stored and the resulting voltage remains available at the op-amp output.

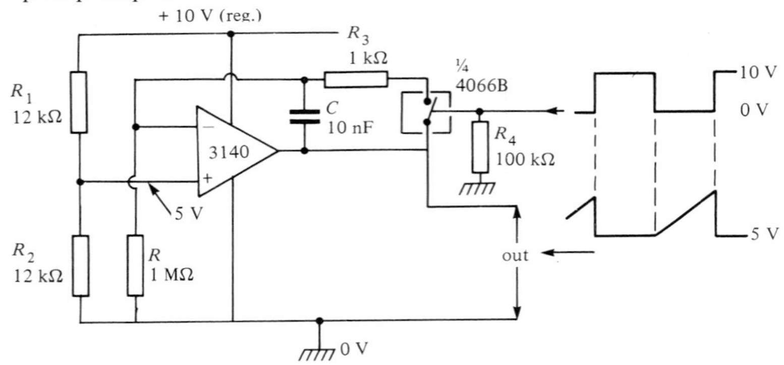

Figure 4.30 *Using the 4066B to implement a ramp generator circuit*

Finally, *Figure 4.30* shows a bilateral switch used in a linear ramp generator. Here, the op-amp is used as an integrator, with its non-inverting pin biased at 5 volts via R_1-R_2, so that a constant current of 5 μA flows into the inverting pin via R. When the bilateral switch is open, this current linearly charges capacitor C, causing a rising ramp to appear at the op-amp output. When the bilateral switch closes, C is rapidly discharged via R_3 and the output switches down to 5 volts.

The basic *Figure 4.30* circuit is quite versatile. The switching can be activated automatically via a free-running oscillator or via a voltage trigger circuit. Bias levels can be shifted by changing the R_1-R_2 values or can be switched automatically via another 4066B stage, enabling a variety of ramp waveforms to be generated.

5 Clock generators

The square-wave generator is one of the most basic circuit blocks used in modern electronics. It can be used for flashing LED indicators, for generating audio and alarm tones, or, if its leading and trailing edges are really sharp, for clocking logic or counter/divider circuitry. Thus a clock generator can be simply described as a high-quality square-wave or trigger-waveform generator.

Practical square-wave and clock generator circuits are easy to design, and can be based on a wide range of semiconductor technologies, including the humble bipolar transistor, the op-amp, the 555 timer IC, and CMOS logic elements. In the present volume, we'll confine the subject entirely to CMOS based designs.

Inexpensive CMOS logic ICs such as the 4001B, 4011B, and 4093B can easily be used to make very inexpensive but highly versatile clock generator circuits. They can be designed to give symmetrical or non-symmetrical outputs, and can be of the free-running or the gated types. In the latter case, they can be designed to turn on with either logic-0 or logic-1 gate signals, and to give either a logic-0 or a logic-1 output when in the off mode. These very inexpensive CMOS circuits can even be used as simple voltage-controlled oscillators (VCOs) or as frequency-modulated oscillators.

If the reader wants really good VCO operation from a square-wave generator, with excellent linearity and versatility, he can turn to the slightly more expensive 4046B phase-locked-loop CMOS IC. We'll look at some practical applications of the 4046B later in this chapter. In the meantime, let's look at some basic CMOS astable clock generator circuits.

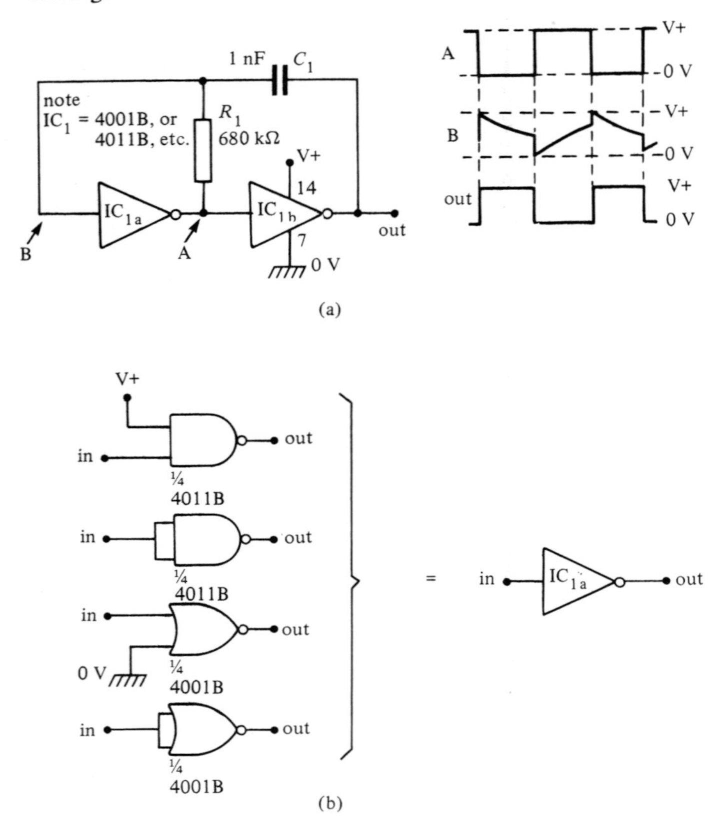

Figure 5.1 *(a) Circuit and waveforms of basic two-stage 1 kHz CMOS astable. (b) Ways of connecting a two-input NAND (4011B) or NOR (4001B) gate as an inverter*

CMOS astable basics

The simplest way to make a CMOS square-wave generator is to wire two CMOS inverter stages in series and use the *C-R* feedback network shown in the basic two-stage astable circuit of *Figure 5.1(a)*. This circuit generates a good square-wave output from IC_{1b} (and a not-quite-so-good anti-phase square-wave output from IC_{1a}), and operates at about 1 kHz with the component values shown. The circuit is suitable for use in many (but not all) clock generator applications, and operates as follows.

In *Figure 5.1(a)* the two inverters are wired in series, so the output of one goes high when the other goes low and vice versa. Time-constant network

C_1-R_1 is wired between the outputs of IC_{1b} and IC_{1a}, with the C_1-R_1 junction fed to the input of the IC_{1a} inverter stage. Suppose initially that C_1 is fully discharged, and that the output of IC_{1b} has just switched high (and the output of IC_{1a} has just switched low).

Under this condition, the C_1-R_1 junction voltage is initially at full positive supply volts, driving the output of IC_{1a} hard low, but this voltage immediately starts to decay exponentially as C_1 charges up via R_1, until eventually it falls into the linear transfer voltage range of IC_{1a}, making its output start to swing high. This swing is amplified by inverter IC_{1b}, initiating a regenerative action in which IC_{1b} output switches abruptly to the low state (and IC_{1a} output switches high). This switching action makes the charge of C_1 try to apply a negative voltage to the input of IC_{1a}, but the built-in protection diodes of IC_{1a} prevent this and instead discharge C_1.

Thus, at the start of the second cycle, C_1 is again fully discharged, so in this case the C_1-R_1 junction is initially at zero volts (driving IC_{1a} output high). However, the voltage then rises exponentially as C_1 charges up via R_1, until eventually it rises into the linear transfer voltage range of IC_{1a}. This initiates another regenerative switching action in which IC_{1b} output switches high again (and IC_{1a} output switches low), and C_1 is initially discharged via the IC_{1a} input protection diodes. The operating cycle then continues *ad infinitum*.

The operating frequency of the above circuit is inversely proportional to the *C*-*R* time constant (the period is roughly 1.4*CR*), and so can be raised by lowering the values of either C_1 or R_1. C_1 must be a non-polarized capacitor and have any value from a few tens of picofarads to several microfarads, and R_1 can have any value from about 4.7 kΩ to 22 MΩ; the astable operating frequency can vary from a fraction of a hertz to about 1 MHz. For variable-frequency operation, wire a fixed and a variable resistor in series in the R_1 position.

Note at this point that each of the inverter stages of the *Figure 5.1(a)* circuit can be made from a single gate of a 4001B quad two-input NOR gate or a 4011B quad two-input NAND gate etc. by using the connections shown in *Figure 5.1(b)*. Thus each of these ICs can provide two astable circuits. Also note that the inputs of all unused gates in these ICs must be tied to one or other of the supply line terminals. The *Figure 5.1(a)* astable (and all other astables shown in this chapter) can be used with any supplies in the range 3 V to 18 V; the zero volts terminal goes to pin 7 of the 4001B or 4011B, and the positive terminal goes to pin 14.

The output of the *Figure 5.1(a)* astable switches (when lightly loaded) almost fully between the zero and positive supply rail values, but the C_1-R_1 junction voltage is prevented from swinging below zero or above the positive supply rail levels by the built-in clamping diodes at the input of IC_{1a}. This factor makes the operating frequency somewhat dependent on supply rail voltages. Typically, the frequency falls by about 0.8 per cent for a 10 per cent rise in supply voltage; if the frequency is normalized with a 10 V supply, the frequency falls by 4 per cent at 15 V or rises by 8 per cent at 5 V.

The operating frequency of the *Figure 5.1(a)* circuit is also influenced by the transfer voltage value of the individual IC_{1a} inverter/gate that is used in the astable, and can be expected to vary by as much as 10 per cent between different ICs. The output symmetry of the square waveform also depends on the transfer voltage value, and in most cases the circuit will give a non-symmetrical output. In most non-precision and hobby applications these defects are, however, of little practical importance.

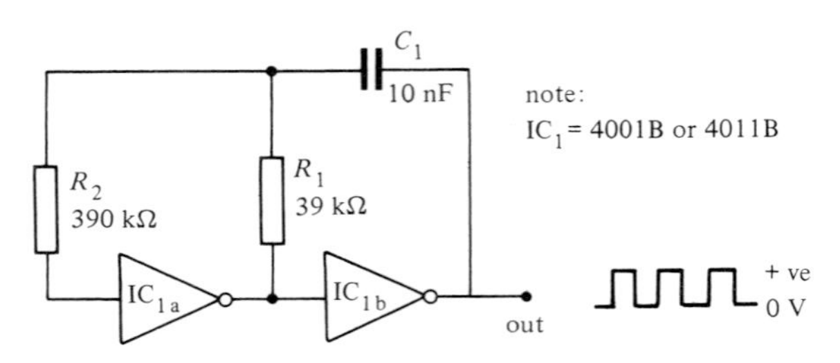

Figure 5.2 *This compensated version of the 1 kHz astable has excellent frequency stability*

Astable variations

Some of the defects of the *Figure 5.1(a)* circuit can be minimized by using the 'compensated' astable of *Figure 5.2* in which R_2 is wired in series with the input of IC_{1a}. This resistor must have a value that is large relative to R_1, and its main purpose is to allow the C_1-R_1 junction to swing freely below the zero and above the positive supply rail voltages during the astable operation and thus improve the frequency stability of the circuit. Typically, when R_2 is ten times the value of R_1, the frequency varies by only 0.5 per cent when the supply voltage is varied between 5 and 15 volts. An incidental benefit of R_2 is that it gives a slight improvement in the symmetry of the astable output waveform.

The basic and compensated astable circuits of *Figures 5.1* and *5.2* can be built with a good number of detail variations, as shown in *Figures 5.3* to *5.6*. In the basic astable circuit, for example, C_1 alternately charges and discharges via R_1 and thus has a fixed symmetry. *Figures 5.3* to *5.5* show how the basic circuit can be modified to give alternate C_1 charge and discharge paths and thus to allow the symmetry to be varied at will.

The *Figure 5.3* circuit generates a highly non-symmetrical waveform, equivalent to a fixed pulse delivered at a fixed timebase rate. Here, C_1 charges in one direction via R_2 in parallel with the D_1-R_1 combination, to generate the

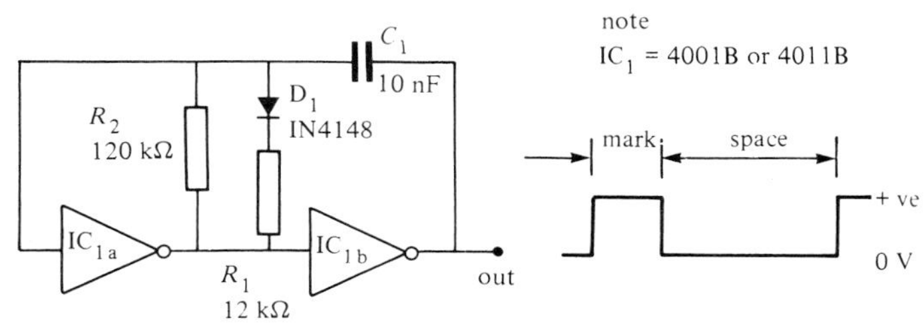

Figure 5.3 *Modifying the astable to give a non-symmetrical output: mark is controlled by the parallel values of* R_1 *and* R_2 *space is controlled by* R_2 *only*

mark or pulse part of the waveform, but discharges in the reverse direction via R_2 only, to give the *space* between the pulses.

Figure 5.4 shows the modifications for generating a waveform with independently variable mark and space times; the mark is controlled by R_1-RV_1-D_1, and the space by R_1-RV_2-D_2.

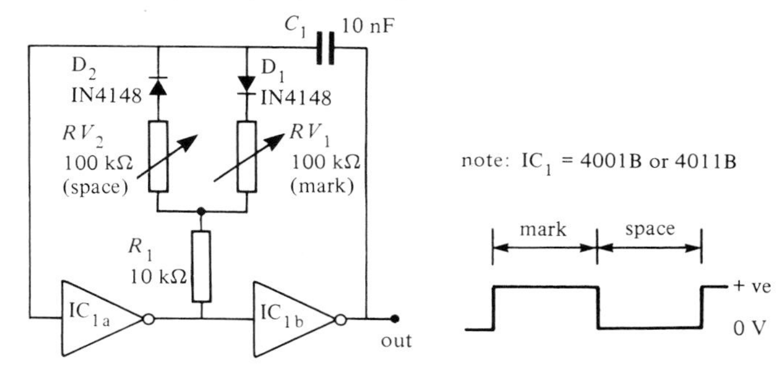

Figure 5.4 *This astable has independently variable mark and space times*

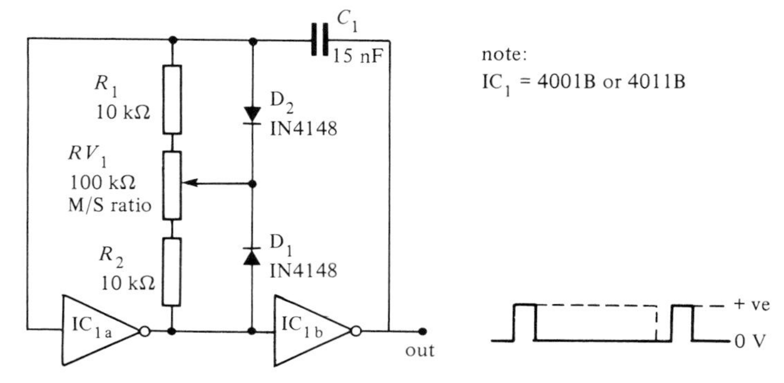

Figure 5.5 *The mark/space ratio of this astable is fully variable from 1:11 to 11:1 via* RV_1*; frequency is almost constant at about 1 kHz*

Figure 5.5 shows the modifications to give a variable-symmetry or M/S ratio output while maintaining a near-constant frequency. Here, C_1 charges in one direction via D_2 and the lower half of RV_1 and R_2, and in the other direction via D_1 and the upper half of RV_1 and R_1. The M/S ratio can be varied over the range 1:11 to 11:1 via RV_1.

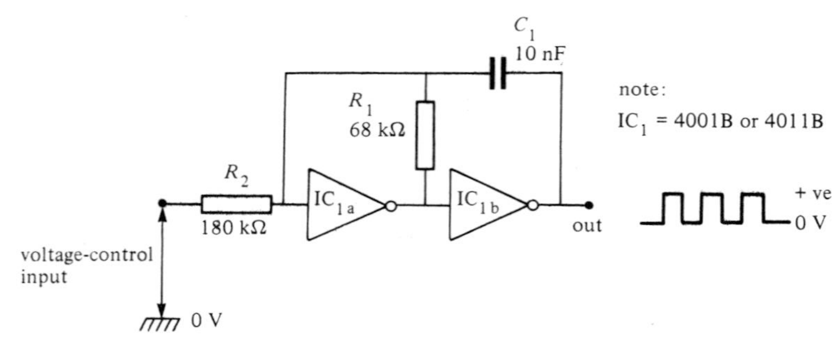

Figure 5.6 *Simple VCO circuit*

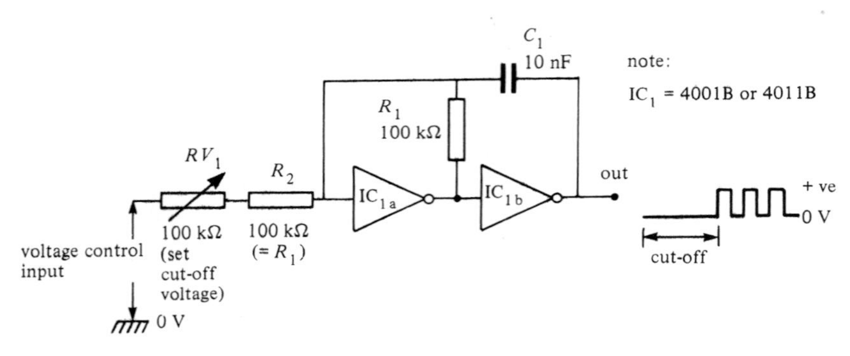

Figure 5.7 *Special-effects VCO which cuts off when* V_{in} *falls below a preset value*

Finally, *Figures 5.6* and *5.7* show a couple of ways of using the basic astable circuit as a very simple VCO. The *Figure 5.6* circuit can be used to vary the operating frequency over a limited range via an external voltage. R_2 must be at least twice as large as R_1 for satisfactory operation, the actual value depending on the required frequency-shift range: a low R_2 value gives a large shift range, and a large R_2 value gives a small shift range. The *Figure 5.7* circuit acts as a special-effects VCO in which the oscillator frequency rises with input voltage, but switches off completely when the input voltage falls below a value preset by RV_1.

Gated astable circuits

All of the astable circuits of *Figure 5.1* to *5.5* can be modified for gated operation, so that they can be turned on and off via an external signal, by simply using a two-input NAND (4011B) or NOR (4001B) gate in place of the inverter in the IC_{1a} position and by applying the input gate control signal to one of the gate input terminals. Note, however, that the 4001B and the 4011B give quite different types of gate control and output operation in these applications, as shown by the two basic versions of the gated astable in *Figures 5.8* and *5.9*.

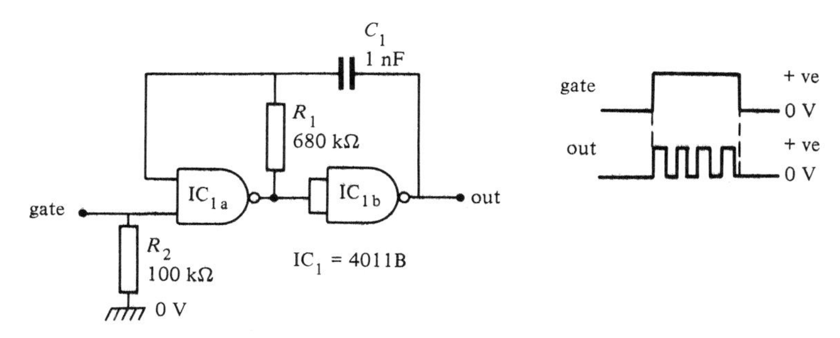

Figure 5.8 *This gated astable has a normally low output and is gated on by a high (logic-1) input*

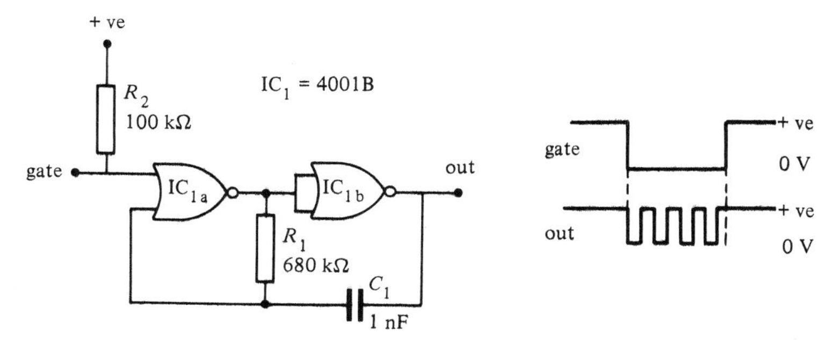

Figure 5.9 *This version of the gated astable has a normally high output and is gated by a low (logic-0) input*

Note specifically from these two circuits that the NAND version is gated on by a logic-1 input and has a normally low output, while the NOR version is gated on by a logic-0 input and has a normally high output. R_2 can be eliminated from these circuits if the gate drive is direct coupled from the output of a preceding CMOS logic stage, for example.

Note that in the basic gated astable circuits of *Figures 5.8* and *5.9* the output signal terminates as soon as the gate drive is removed; consequently, any noise present at the gate terminal also appears at the outputs of these circuits. *Figures 5.10* and *5.11* show how to modify the circuits so that they produce noiseless outputs.

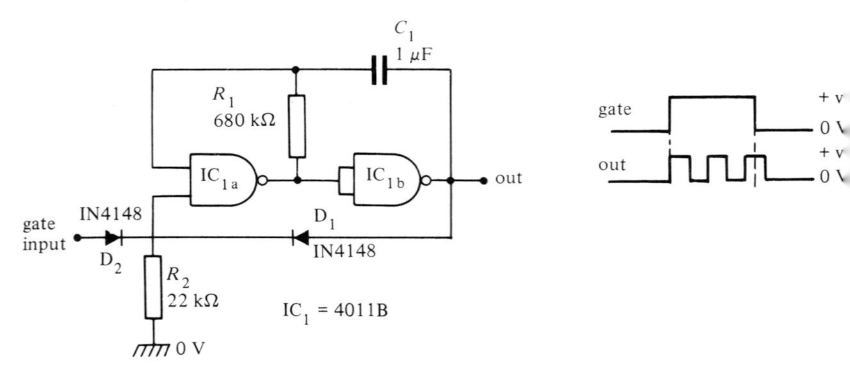

Figure 5.10 *Semi-latching or noiseless gated astable circuit, with logic-1 gate input and normally zero output*

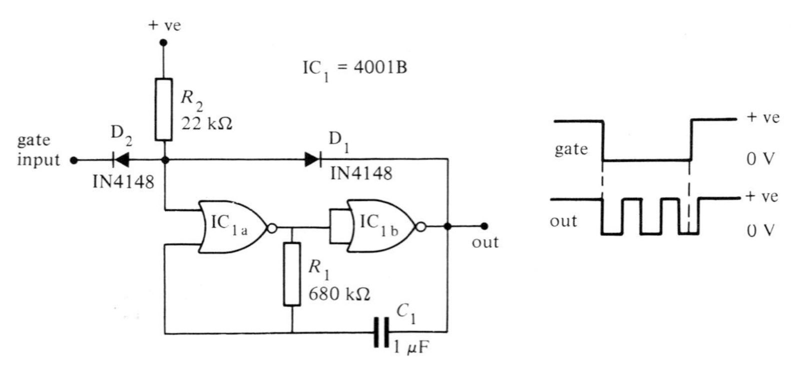

Figure 5.11 *Alternative semi-latching gated astable, with logic-0 gate input and normally high output*

Here, the gate signal of IC_{1a} is derived from both the outside world and from the output of IC_{1b} via diode OR gate D_1-D_2-R_2. As soon as the circuit is gated from the outside world via D_2 the output of IC_{1b} reinforces or self-latches the gating via D_1 for the duration of one half astable cycle, thus eliminating any effects of a noisy outside world signal. The outputs of these semi-latching gated astable circuits are thus always complete numbers of half cycles.

Ring-of-three-astable

The two-stage astable circuit is a good general-purpose square-wave generator. However, it is not always suitable for direct use as a clock generator with fast-acting counting and dividing circuits, since it tends to pick up and amplify any existing supply line noise during the 'transitioning' parts of its operating cycle and to thus produce output square waves with distorted leading and trailing edges. A far better type of clock generator circuit is the ring-of-three astable shown in *Figure 5.12*.

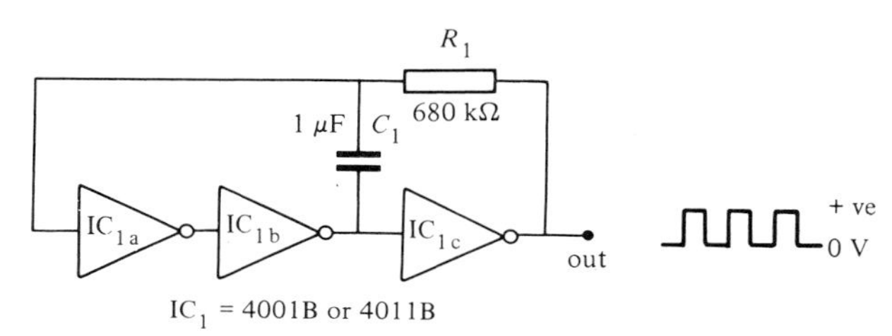

Figure 5.12 *This ring-of-three astable makes an excellent clock generator*

The *Figure 5.12* ring-of-three circuit is similar to the basic two-stage astable, except that its input stage (IC_{1a}-IC_{1b}) acts as an ultra-high-gain non-inverting amplifier and its main timing components (C_1-R_1) are transposed (relative to the two-stage astable). Because of the very high overall gain of the circuit, it produces an excellent and distortion-free square-wave output, ideal for clock generator use.

The basic ring-of-three astable can be subjected to all the design modifications that we've already looked at for the basic two-stage astable, e.g.

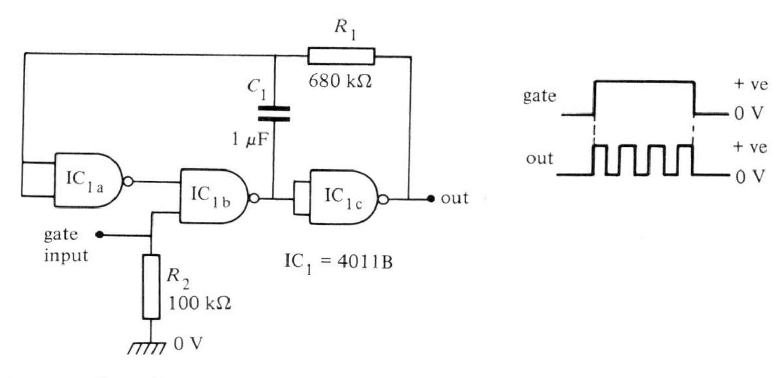

Figure 5.13 *This gated ring-of-three astable is gated by a logic-1 input and has a normally low output*

it can be used in either basic or compensated form and can give either a symmetrical or a non-symmetrical output. The most interesting variations of the circuit occur, however, when it is used in the gated mode, since it can be gated via either the IC_{1b} or the IC_{1c} stages. *Figures 5.13* to *5.16* show four variations on this gating theme.

Thus the *Figure 5.13* and *5.14* circuits are both gated on by a logic-1 input signal, but the *Figure 5.13* circuit has a normally low output whereas that of *Figure 5.14* is normally high. Similarly the *Figure 5.15* and *5.16* circuits are both gated on by a logic-0 signal, but the output of the *Figure 5.15* circuit is normally low whereas that of *Figure 5.16* is normally high.

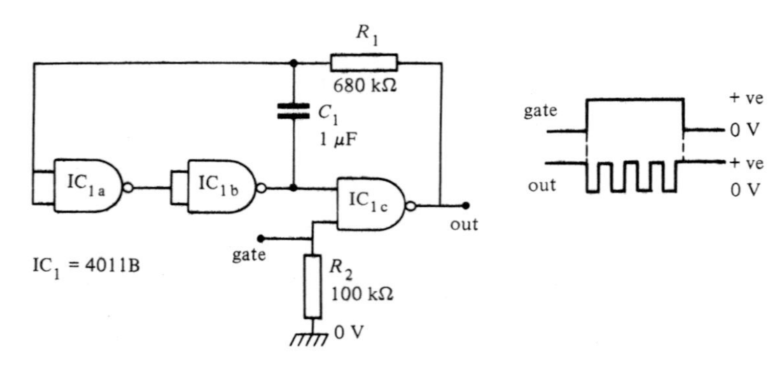

Figure 5.14 *This gated ring-of-three astable is gated by a logic-1 input and has a normally high output*

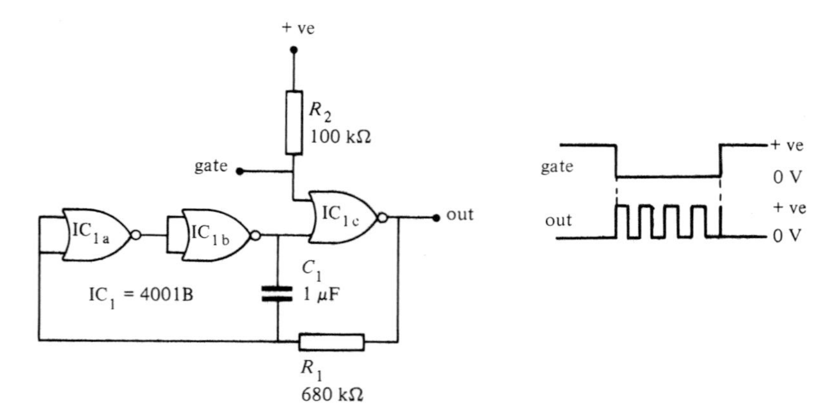

Figure 5.15 *This gated ring-of-three astable is gated by a logic-0 input and has a normally low output*

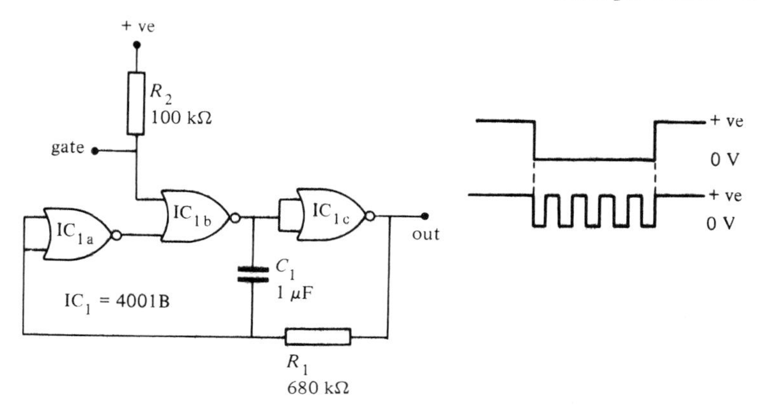

Figure 5.16 *This gated ring-of-three astable is gated by a logic-0 input and has a normally high output*

The Schmitt astable

An excellent astable clock generator can also be made from a single CMOS Schmitt inverter stage. Suitable ICs for use in this application are the 40106B hex Schmitt inverter, and the 4093B quad two-input NAND Schmitt trigger (see *Figure 3.26*). In the latter case, each NAND gate of the 4093B can be used as an inverter by simply disabling one of its input terminals, as shown in the basic Schmitt astable circuit of *Figure 5.17*.

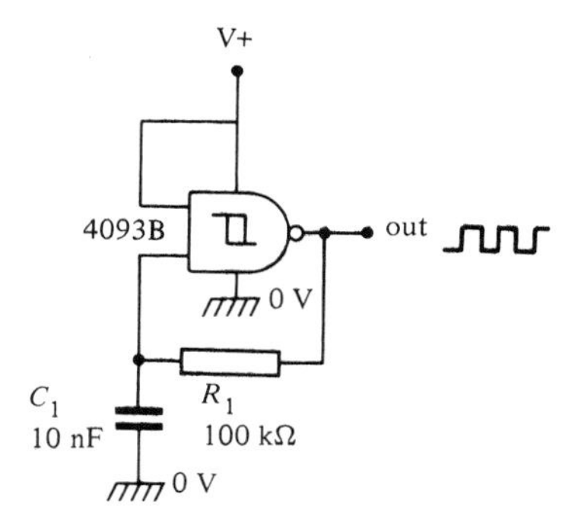

Figure 5.17 *Basic Schmitt astable*

The Schmitt astable circuit gives an excellent performance, with very clean output edges that are unaffected by supply line ripple and other perturbations. The operating frequency is determined by the C_1-R_1 values, and can be varied from a few cycles per minute to 1 MHz or so. The circuit action is

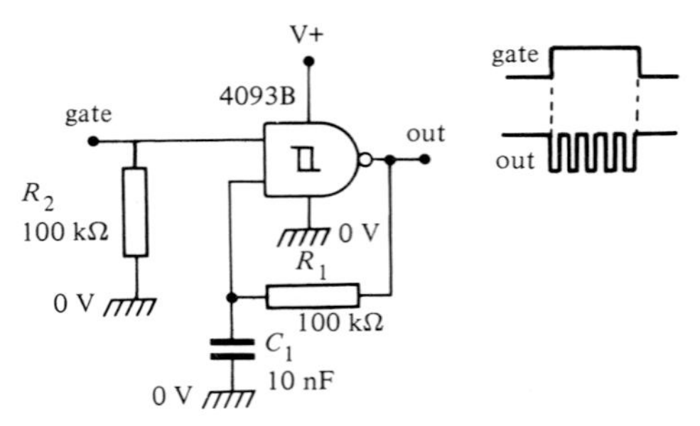

Figure 5.18 *Gated Schmitt astable*

such that C_1 alternately charges and discharges via R_1, without switching the C_1 polarity; C_1 can thus be a non-polarized component.

Figure 5.18 shows how the above 4093B-based astable circuit can be modified so that it can be gated on and off via an external signal. Note that the circuit is gated on by a high (logic-1) input, but gives a high output when it is in the gated-off state.

The basic astable circuit of *Figure 5.17* generates an inherently symmetrical square-wave output. The circuit can be made to produce a non-symmetrical output by providing its timing capacitor with alternate charge and discharge paths, as shown in the circuits of *Figures 5.19* and *5.20*. The *Figure 5.19* circuit produces a fixed M/S ratio output. The M/S ratio of the *Figure 5.20* circuit can be varied over a wide range via RV_1.

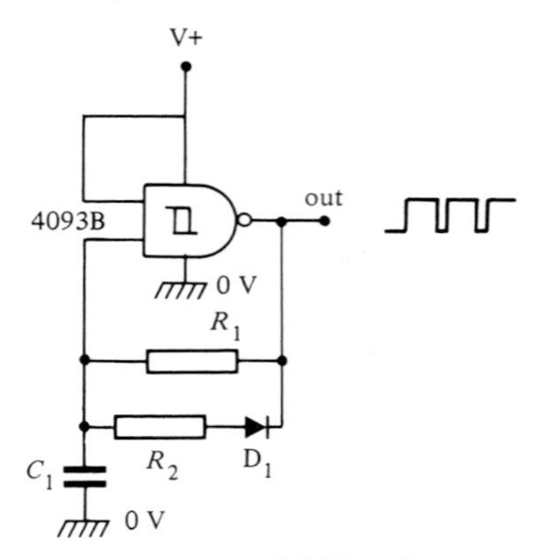

Figure 5.19 *Astable with non-symmetrical M/S ratio*

A CMOS 555 IC

Most readers will know that the popular 555 timer IC can be wired in the astable mode and used to generate excellent square-wave output signals. They

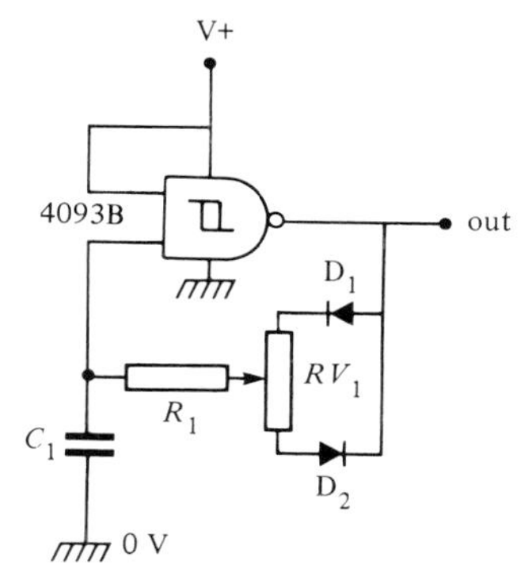

Figure 5.20 *Astable with variable M/S ratio*

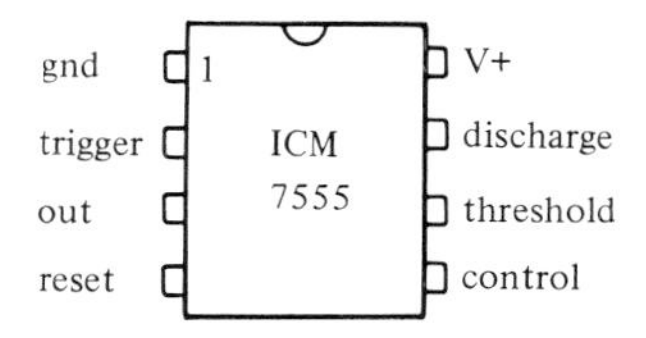

Figure 5.21 *Outline of the ICM7555 CMOS 555 IC*

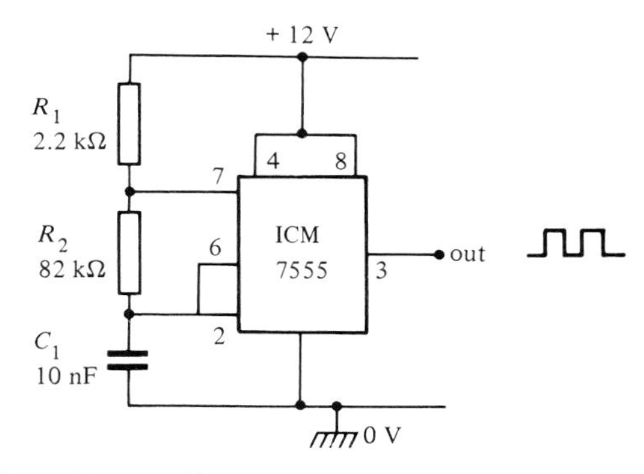

Figure 5.22 *Basic 7555 astable*

may not know that a CMOS version of this device is also available, and is known as the ICM7555. *Figure 5.21* shows the outline of this IC and *Figure 5.22* shows how it can be used in the basic astable mode.

The action of the *Figure 5.22* basic astable is such that C_1 alternately charges via R_1-R_2 and discharges via R_2 only, generating an excellent near-symmetrical square-wave output that is suitable for use as a clock waveform. Note that R_2 acts as the main time-constant resistor and can have any value from about 4 kΩ to 22 MΩ; C_1 is the time-constant capacitor and can have any value from a few picofarads to many hundreds of microfarads, and may be of the polarized or non-polarized types.

The astable circuit can only work if pin 4 is tied to the positive supply rail; if pin 4 is grounded, the astable is disabled. The astable can thus be used in the gated mode by simply wiring pin 4 as shown in *Figure 5.23*.

The basic astable circuit generates an almost symmetrical output waveform. It can be made to generate a non-symmetrical waveform in a variety of ways. *Figure 5.24* shows one useful variation. In this case C_1

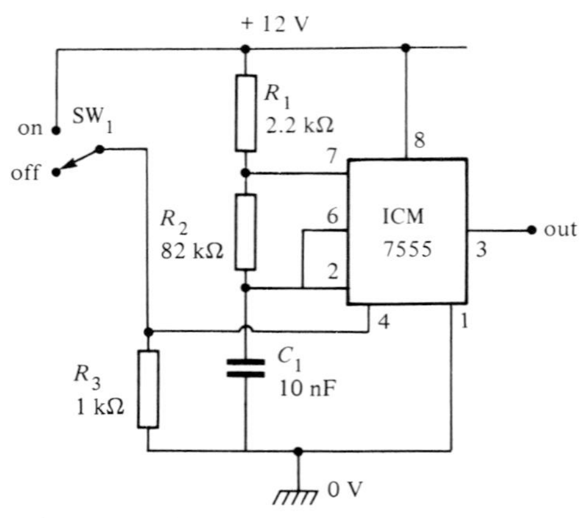

Figure 5.23 *Gated 7555 astable*

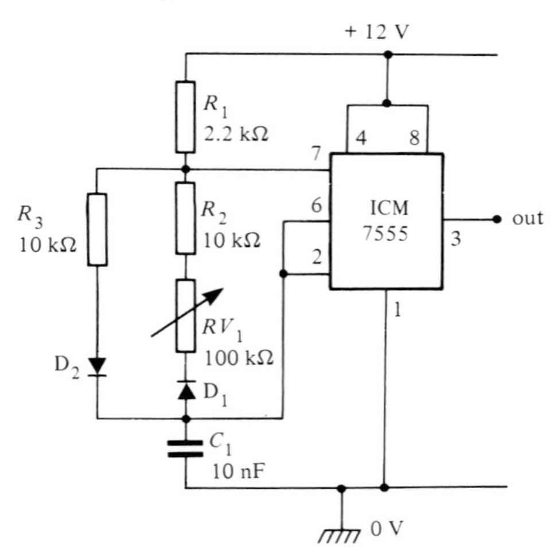

Figure 5.24 *Astable with variable M/S ratio*

alternately charges via R_1-R_3 and D_2, and discharges via D_1-RV_1 and R_2; the output waveform symmetry of this circuit is thus fully variable via RV_1.

4046B VCO circuits

To complete this look at CMOS square-wave and clock generator circuits, let's consider some practical VCO (voltage-controlled oscillator) applications of the 4046B phase-locked-loop (PLL) IC. *Figure 5.25* shows the internal block diagram and pin-outs of this chip, which contains a couple of phase comparators, a VCO, a zener diode, and a few other bits and pieces.

For our present purpose, the most important part of the chip is the VCO section. This is a highly versatile device. It produces a well-shaped symmetrical square-wave output, has a top-end frequency limit in excess of 1 MHz, has a voltage-to-frequency linearity of about 1 per cent and can easily be scanned

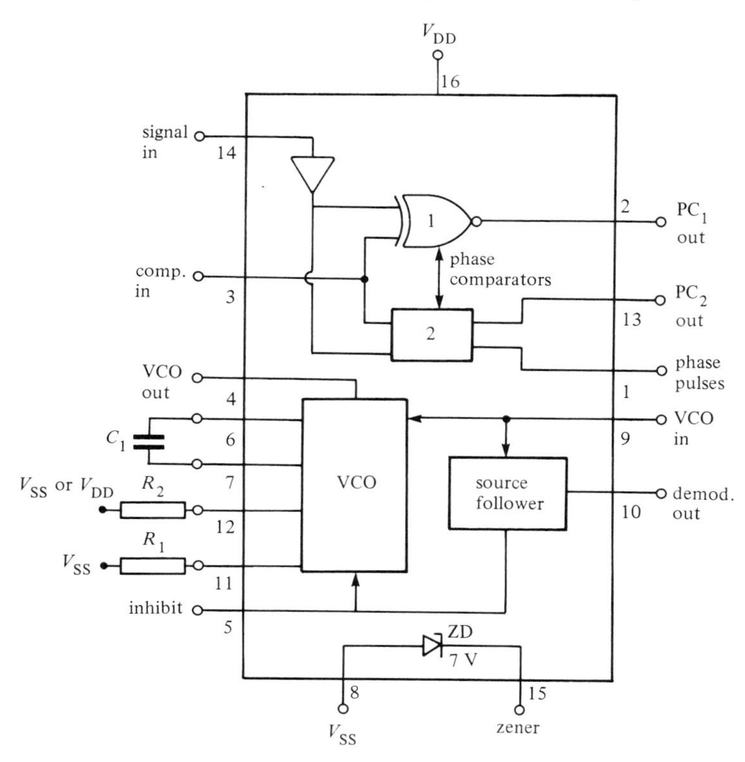

Figure 5.25 *Internal block diagram and pin-outs of the 4046B*

through a 1 000 000:1 range by an external voltage applied to the VCO input terminal. The frequency of the oscillator is governed by the value of a capacitor (minimum value 50 pF) connected between pins 6 and 7, by the value of a resistor (minimum value 10 kΩ) wired between pin 11 and ground, and by the voltage (any value from zero to the supply voltage value) applied to VCO input pin 9.

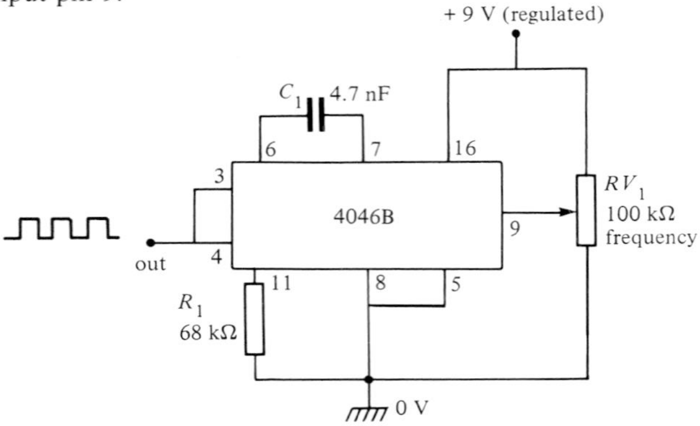

Figure 5.26 *Basic wide-range VCO, spanning near-zero to roughly 5 kHz via* RV_1.

Figure 5.26 shows the simplest possible way of using the 4046B VCO as a voltage-controlled square-wave generator. Here, C_1-R_1 determines the maximum frequency that can be obtained (with the pin 9 voltage at maximum) and RV_1 controls the actual frequency by applying a control voltage to pin 9: the frequency falls to a very low value (a fraction of a hertz)

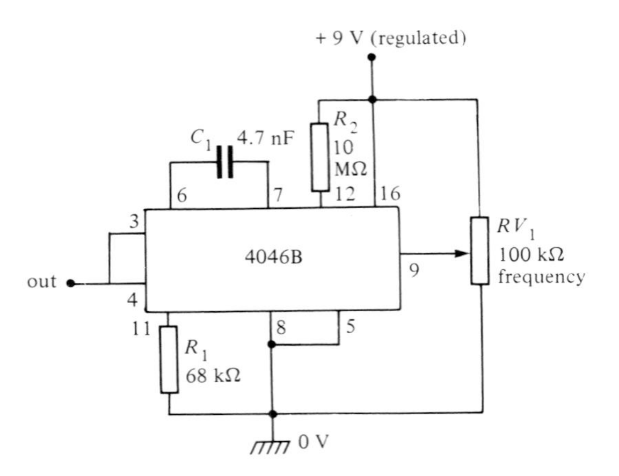

Figure 5.27 *The frequency of this VCO is variable all the way down to zero*

with pin 9 at zero volts. The effective voltage-control range of pin 9 varies from roughly 1 V below the supply value to about 1 V above zero, and gives a frequency span of about 1 000 000:1. Ideally, the circuit supply voltage should be regulated.

It was stated above that the frequency falls to near-zero when the input voltage of the *Figure 5.26* circuit is reduced to zero. *Figure 5.27* shows how the circuit can be modified so that the frequency falls all the way to zero with zero input, by wiring high-value resistor R_2 between pins 12 and 16. Note here that, when the frequency is reduced to zero, the VCO output randomly settles in either a logic-0 or a logic-1 state.

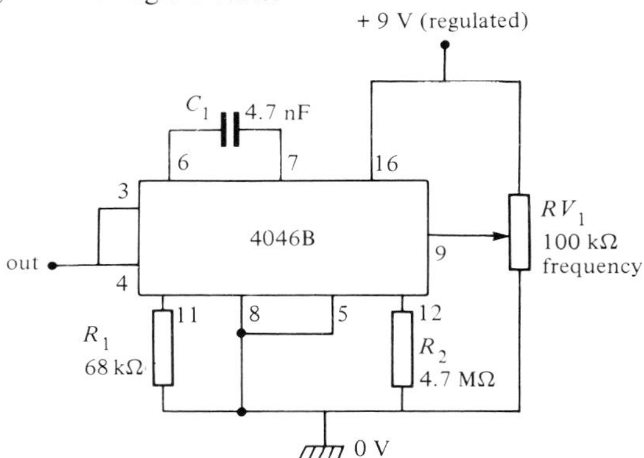

Figure 5.28 *Restricted-range VCO, with frequency variable from roughly 72 Hz to 5 kHz via* RV_1.

Figure 5.28 shows how the pin 12 resistor can alternatively be used to determine the minimum operating frequency of a restricted-range VCO. Here, f_{min} is determined by C_1-R_2, and f_{max} is determined by C_1 and the parallel resistance of R_1 and R_2.

Figure 5.29 shows an alternative version of the restricted-range VCO in which f_{max} is controlled by C_1-R_1, and f_{min} is determined by C_1 and the series combination of R_1 and R_2. Note that, by suitable choice of the R_1 and R_2 values, the circuit can be made to span any desired frequency range from 1:1 to near-infinity.

Finally, it should be noted that the VCO section of the 4046B can be disabled by taking pin 5 of the package high (to logic-1) or enabled by taking pin 5 low (to logic-0). This feature makes it possible to gate the VCO on and off via external signals. Thus, *Figure 5.30* shows how the basic VCO circuit can be gated via a signal applied to an external inverter stage. Alternatively, *Figure 5.31* shows how one of the internal phase comparators of the 4046B can be used to provide gate inversion, so that the VCO can be gated via an external voltage applied to pin 3.

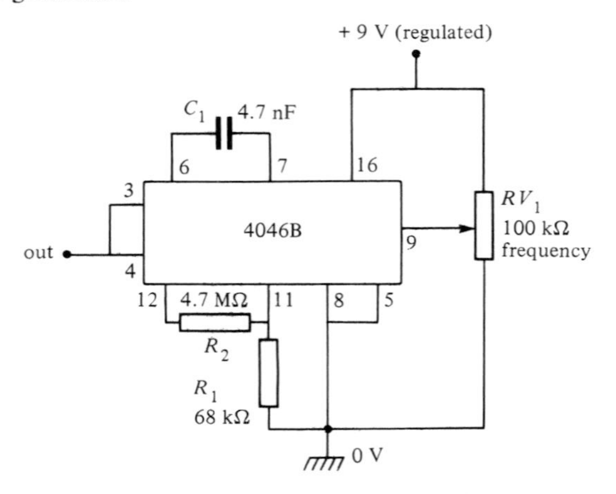

Figure 5.29 *Alternative version of the restricted-range VCO.* f_{max} *is controlled by* C_1-R_1, f_{min} *by* C_1-$(R_1 + R_2)$

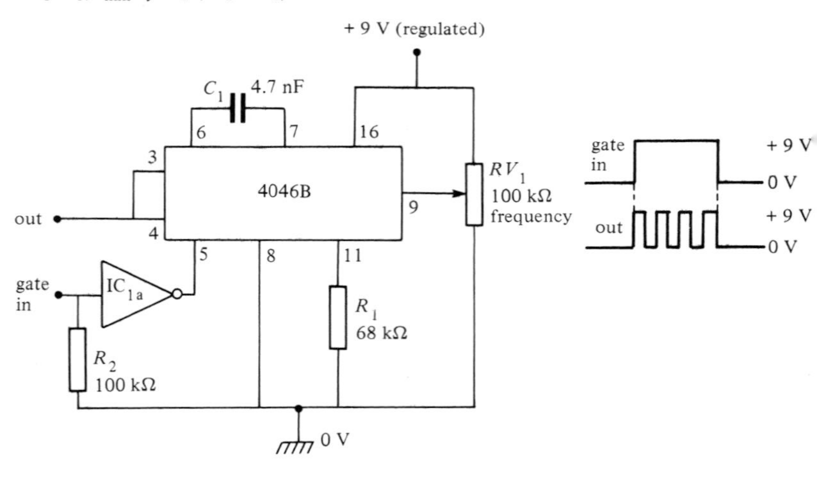

Figure 5.30 *Gated wide-range VCO, using an external gate inverter*

Bistable circuits

Strictly speaking, a clock generator can be any circuit that generates a clean (noiseless, with sharp leading and trailing edges) waveform suitable for clocking modern fast digital counter/divider circuitry. Thus a clock generator can take the form of a simple Schmitt trigger that converts a sine-wave input into a good square-wave output, or a square-wave generator of the type

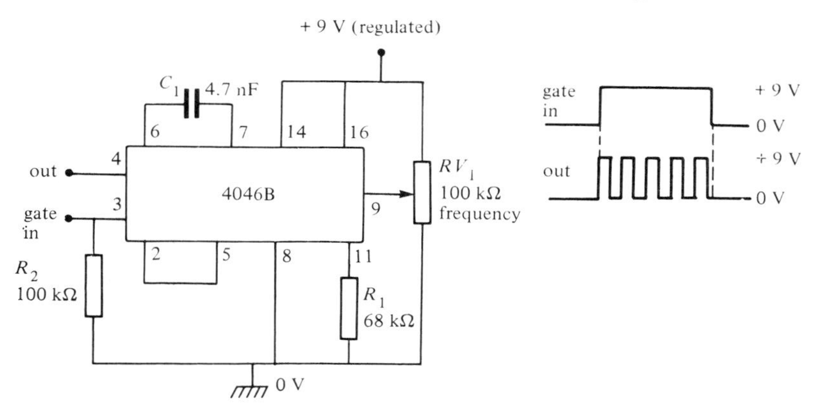

Figure 5.31 *Gated wide-range VCO, using one of the internal phase comparators as a gate inverter*

already described in this chapter, or of a monostable pulse generator of the type described in chapter 6, and so on.

One particularly useful type of clock generator is the simple R-S (reset-set) bistable (also known as the R-S flip-flop), which can be used to deliver a single clock pulse each time the available one of its two input terminals is activated, either electronically or via a push-button switch. *Figures 5.32* to *5.35* show four ways of making such bistables, using pairs of 4001B or 4011B gates.

The bistable circuit of *Figure 5.32* is triggered by positive-going input pulses, and is designed around 4001B NOR-type gates. *Figure 5.33* shows how the circuit can be modified for push-button triggering.

The bistable circuit of *Figure 5.34* is triggered by negative-going input pulses, and is designed around 4011B NAND-type gates. *Figure 5.35* shows the circuit modified for push-button triggering.

Note that, in all the above circuits, once the bistable has been set its state

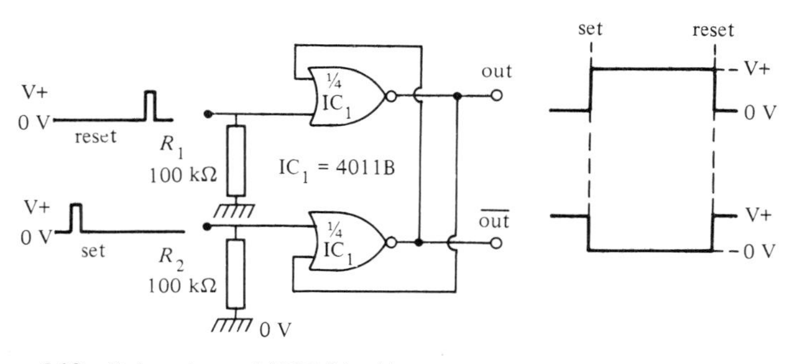

Figure 5.32 *Pulse-triggered NOR bistable*

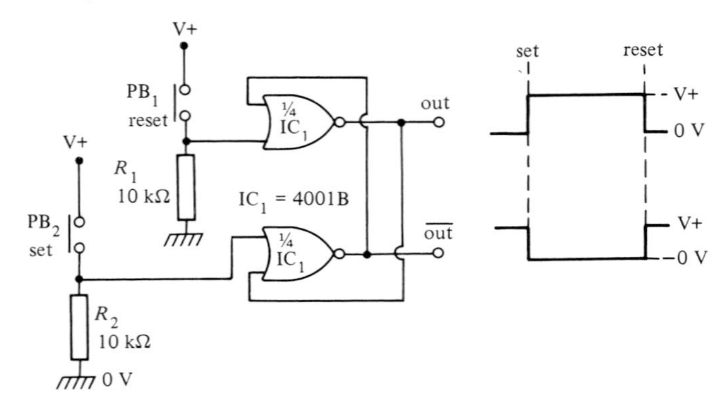

Figure 5.33 *Manually triggered NOR bistable*

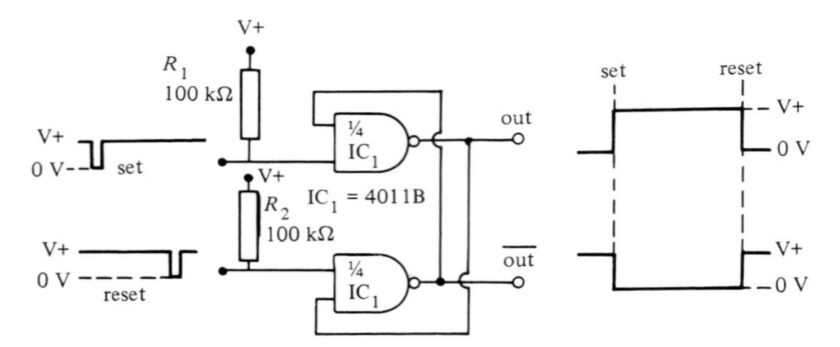

Figure 5.34 *Pulse-triggered NAND bistable*

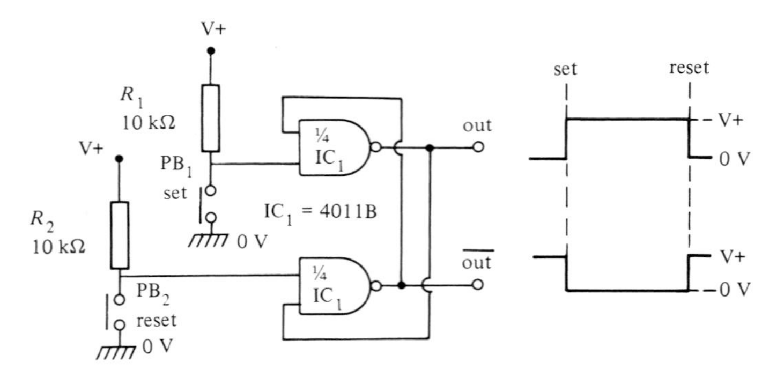

Figure 5.35 *Manually triggered NAND bistable*

can not be altered until the reset terminal is activated, and once it has been reset its state cannot be altered until the set terminal is activated. Also note that each circuit provides a pair of anti-phase outputs. Finally, note that other types of flip-flop circuits are discussed in depth in Chapter 7.

6 Pulse generator circuits

In digital electronics the two most commonly used types of waveform are the free-running or gated square wave, and the triggered or one-shot pulse waveform. We dealt with square-wave generator circuits in Chapter 5. This chapter deals exclusively with pulse generator circuits.

Pulse generator basics

Professional and amateur circuit designers often have to devise means of generating pulse waveforms in various parts of a circuit. Sometimes they simply need to generate a pulse of non-critical width on the arrival of the leading or trailing edge of an input square wave, as shown in *Figure 6.1*, and in such cases they may use a circuit element known as a half-monostable or edge detector. At other times they may need to generate a pulse of precise width on the arrival of a suitable trigger signal, and in such cases they may use a standard monostable or one-shot multivibrator circuit.

In the standard monostable circuit, the arrival of the trigger signal initiates an internal timing cycle which causes the monostable output to change state at the start of the timing cycle, but to revert back to its original state on completion of the cycle, as shown in *Figure 6.2*.

Note that once a timing cycle has been initiated the standard monostable circuit is immune to the effects of subsequent trigger signals until its timing period ends naturally. This type of circuit can sometimes be modified by adding a reset control terminal, as shown in *Figure 6.3*, to enable the output pulse to be terminated or aborted at any time via a suitable command signal.

A third type of monostable circuit (mono) is the retriggerable mono. Here

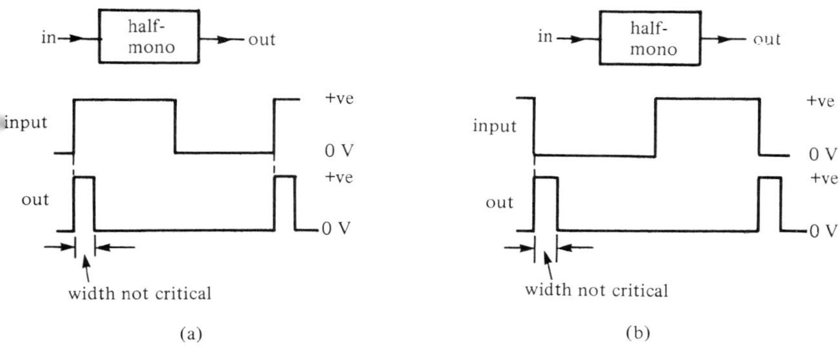

Figure 6.1 *A half-mono circuit may be used to detect (a) the leading or (b) the trailing edge of an input waveform*

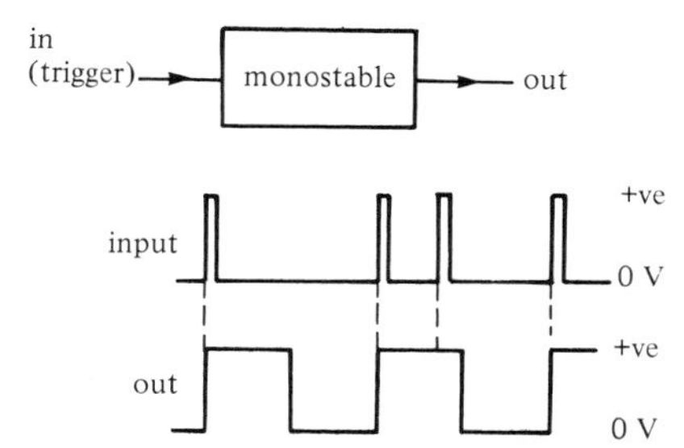

Figure 6.2 *A standard monostable generates an accurate output pulse on the arrival of a suitable trigger signal*

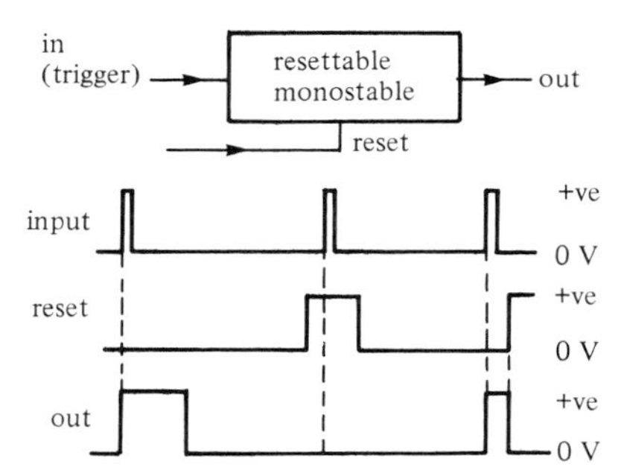

Figure 6.3 *The output pulse of a resettable mono can be aborted by a suitable reset pulse*

the trigger signal actually resets the mono and then, after a very brief delay, initiates a new pulse-generating timing cycle, as shown in *Figure 6.4*. Thus each new trigger signal initiates a new timing cycle, even if the trigger signal arrives during an existing cycle.

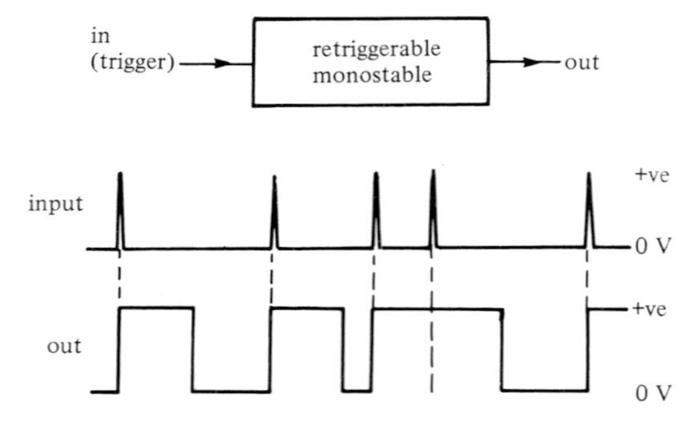

Figure 6.4 *A retriggerable mono starts a new timing cycle on the arrival of each new trigger signal*

Thus the circuit designer may use a half-mono, a standard mono, a resettable mono, or a retriggerable mono to generate pulses in a circuit, the type decision depending on the specific circuit design requirements.

The choice of IC to use in a pulse generator design is usually dictated by considerations of economics and convenience, rather than by the actual design requirements. Thus, if the designer needs a standard CMOS monostable of only modest precision, he can build it very inexpensively by using a logic IC such as a 4001B or 4011B, or can build a circuit of better precision (but greater cost) by using a 7555 timer chip or a dedicated monostable IC such as a 4047B. In this chapter we take a detailed look at CMOS versions of the various types of pulse generator circuit, and at a variety of ways of implementing them.

Edge detector circuits

Edge detector circuits are used to generate an output pulse on the arrival of either the leading or the trailing edge of a rectangular input waveform. In most practical applications the precise width of the output pulse is non-critical. The basic method of making an edge detector is to feed the rectangular input waveform to a short-time-constant *C-R* differentiation network, to produce a double-spike output waveform with sharp leading edges on the arrival of each input edge, and then to eliminate the unwanted part of the spike waveform with a discriminator diode. The remaining spike or sawtooth waveform is then converted into a clean pulse shape by feeding it through a Schmitt trigger or similar circuit. The Schmitt may be of either the inverting or the non-inverting type, depending on the required polarity of the output pulse waveform; in either case, the resulting circuit is known as a half-monostable or half-mono pulse generator.

CMOS Schmitt trigger ICs incorporate built-in protection diodes on all input terminals, and these can be used to perform the discriminator diode action described above. Note that each gate of the popular 4093B quad two-input NOR Schmitt IC (see *Figure 3.26*) can be used as a normal Schmitt inverter by wiring one input terminal to the positive supply rail and using the other terminal as the input point, as shown in *Figure 6.5*. A non-inverting Schmitt can be made by wiring two inverting Schmitts in series, as shown in *Figure 6.6*.

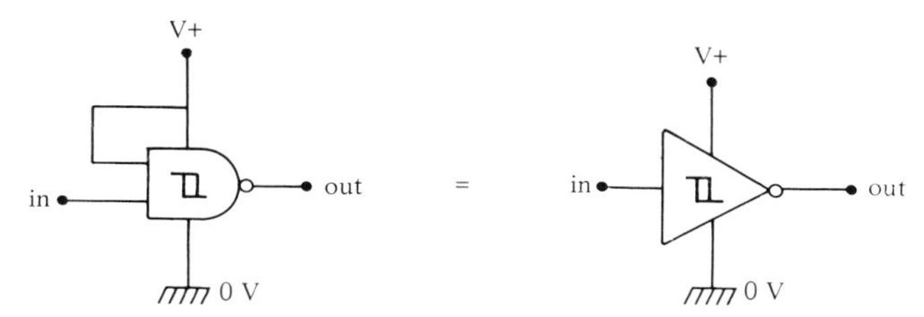

Figure 6.5 *A 4093B two-input NOR Schmitt can be used as a normal Schmitt inverter by wiring one input high*

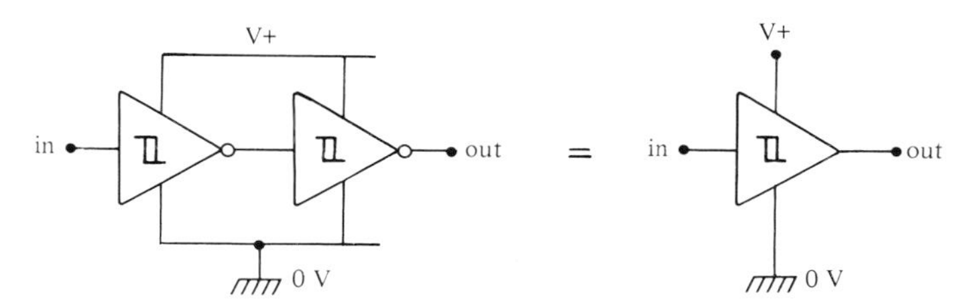

Figure 6.6 *A non-inverting Schmitt can be made by wiring two inverting Schmitt elements in series*

Figure 6.7 shows two ways of making a leading-edge-detecting half-mono circuit. Here, the input of the Schmitt is tied to ground via resistor R, and C-R has a time constant that is short relative to the period of the input waveform. The leading edge of the input signal is thus converted into the spike waveform shown, and this spike is then converted into a good clean pulse waveform via the Schmitt. The circuit generates a positive-going output pulse if a non-inverting Schmitt is used (*Figure 6.7(a)*), or a negative-going output pulse if an inverting Schmitt is used (*Figure 6.7(b)*). In either case, the output pulse has a period P of roughly $0.7CR$.

Figure 6.8 shows how to make a trailing-edge-detecting half-mono. In this case the Schmitt input is tied to the positive supply rail via R, and C-R again has a short time constant. The circuit generates a positive-going output pulse if an inverting Schmitt is used (*Figure 6.8(a)*), or a negative-going pulse if a

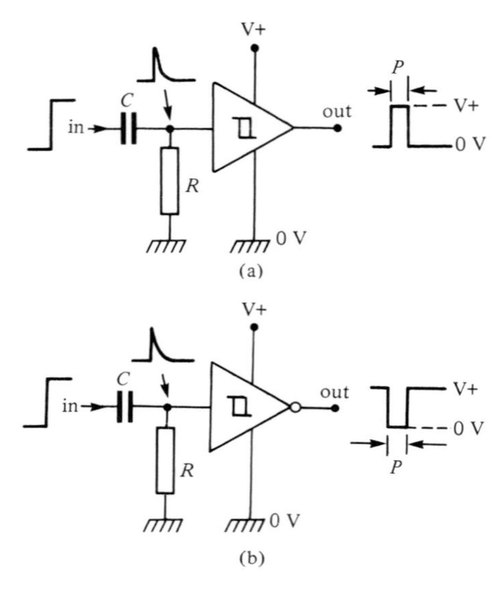

Figure 6.7 *Leading-edge detector circuits giving (a) positive and (b) negative output pulses*

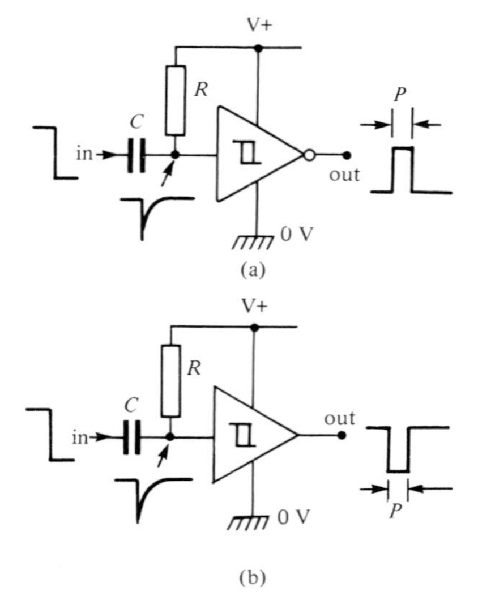

Figure 6.8 *Trailing-edge detector circuit giving (a) positive and (b) negative output pulses*

non-inverting Schmitt is used. The output pulse has a period of roughly $0.7CR$.

Two useful variants of the edge-detecting half-mono circuit are the noiseless push-button switch of *Figure 6.9*, which effectively eliminates the adverse effects of switch contact bounce and noise, and the power-on reset-pulse generator circuit of *Figure 6.10*, which generates a reset pulse when power is first applied to the circuit.

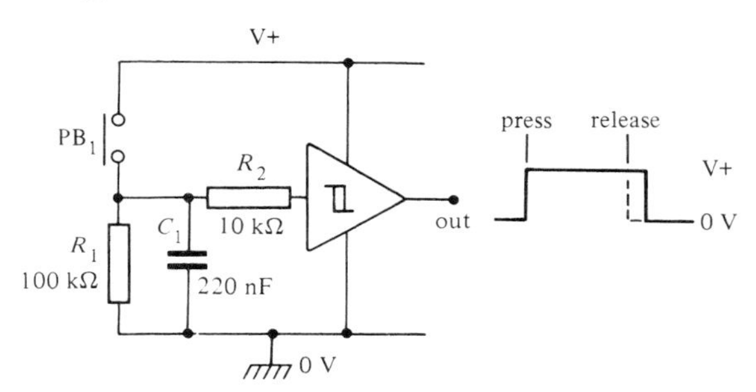

Figure 6.9 *Noiseless push-button switch*

In *Figure 6.9* the input of the non-inverting Schmitt is tied to ground via high-value timing resistor R_1 and by input protection resistor R_2, so the output of the circuit is normally low. When push-button switch PB_1 is closed, C_1 charges rapidly to the full positive supply value, driving the Schmitt output high, but when PB_1 is released again C_1 discharges relatively slowly via R_1, and the Schmitt output does not return low until roughly 20 ms later. The circuit thus ignores the transient switching effects of PB_1 noise and contact bounce etc., and generates a clean output pulse waveform with a period that is roughly 20 ms longer than the mean duration of the PB_1 switch closure.

The *Figure 6.10* circuit uses an inverting Schmitt and produces a 700 ms

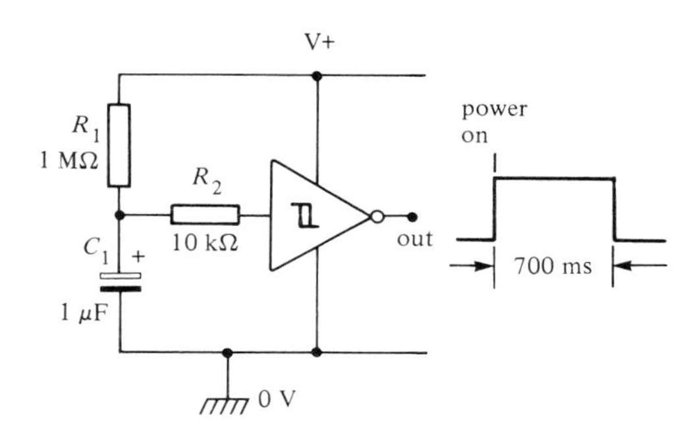

Figure 6.10 *Power-on reset-pulse generator*

output pulse (suitable for resetting external circuitry etc.) when power is first connected. When power is initially connected C_1 is fully discharged, so the Schmitt input is pulled low and its output is switched high; C_1 then charges via R_1 until, after about 700 ms, the C_1 voltage rises to such a level that the Schmitt output switches low, completing the switch-on output pulse.

4001B/4011B monostables

The cheapest way of making a standard or a resettable monostable (or one-shot) pulse generator is to use a 4001B quad two-input NOR gate or a 4011B quad two-input NAND gate in one of the configurations shown in *Figures 6.11* to *6.14*. Note, however, that the output pulse widths of these circuits are subject to fairly large variations between individual ICs and with variations in supply rail voltage, and these circuits are thus not suitable for use in high-precision applications.

Figures 6.11 and *6.12* show alternative versions of the standard monostable

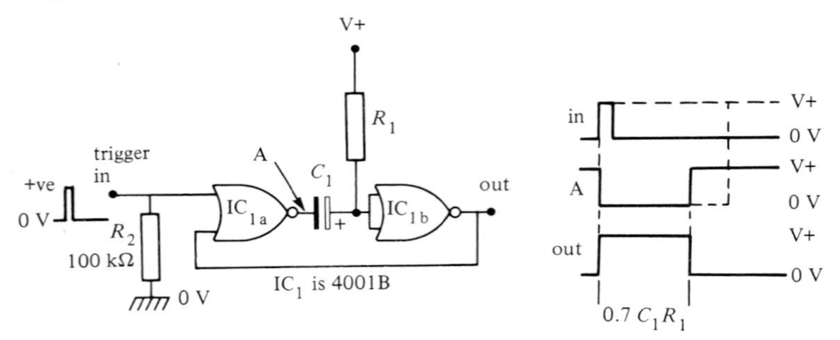

Figure 6.11 *Two-gate NOR monostable is triggered by a positive-going signal and generates a positive-going output pulse*

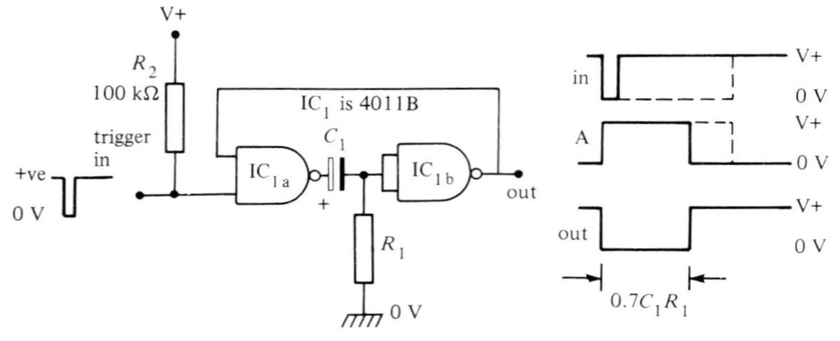

Figure 6.12 *Two-gate NAND monostable is triggered by a negative-going signal and generates a negative-going output pulse*

circuit, each using only two of the four available gates in the specified CMOS package. In these circuits the duration of the output pulse is determined by the C_1-R_1 values, and approximates 0.7 C_1R_1. Thus, when R_1 has a value of 1.5 MΩ the pulse period approximates one second per microfarad of C_1 value. In practice, C_1 can have any value from about 100 pF to a few thousand microfarads, and R_1 can have any value from 4 kΩ to 10 MΩ.

An outstanding feature of these circuits is that the input trigger pulse or signal can be direct coupled and its duration has little effect on the length of the generated output pulse. The NOR version of the circuit (*Figure 6.11*) has a normally low output and is triggered by the edge of a positive-going input signal, and the NAND version (*Figure 6.12*) has a normally high output and is triggered by the edge of a negative-going input signal.

Another feature is that the pulse signal appearing at point A has a period equal to that of either the output pulse or the input trigger pulse, whichever is the greater of the two. This feature is of value when making pulse-length comparators, over-speed alarms, etc.

The operating principle of these monostable circuits is fairly simple. Look first at the case of the *Figure 6.11* circuit, in which IC_{1a} is wired as a NOR gate and IC_{1b} is wired as an inverter. When this circuit is in the quiescent state the trigger input terminal is held low by R_2, and the output of IC_{1b} is also low. Thus both inputs of IC_{1a} are low, so IC_{1a} output is forced high and C_1 is discharged.

When a positive trigger signal is applied to the circuit the output of IC_{1a} is immediately forced low and (since C_1 is discharged at this moment) pulls IC_{1b} input low and thus drives IC_{1b} output high. IC_{1b} output is coupled back to the IC_{1a} input, however, and thus forces IC_{1a} output to remain low irrespective of the prevailing state of the trigger signal. As soon as IC_{1a} output switches low, C_1 starts to charge up via R_1 and, after a delay determined by the C-$_1$-R_1 values, the C_1 voltage rises to such a level that the output of IC_{1b} starts to swing low, terminating the output pulse. If the trigger signal is still high at this moment, the pulse terminates non-regeneratively, but if the trigger signal is low (absent) at this moment the pulse terminates regeneratively.

The *Figure 6.12* circuit operates in a manner similar to that described above, except that IC_{1a} is wired as a NAND gate, with its trigger input terminal tied to the positive supply rail via R_2, and the R_1 timing resistor is taken to ground.

In the *Figure 6.11* and *6.12* circuits the output is direct coupled back to one input of IC_{1a} to effectively maintain a trigger input once the true trigger signal is removed, thereby giving a semi-latching circuit operation. These circuits can be modified so that they act as resettable monostables by simply providing them with a means of breaking this feedback path, as shown in *Figures 6.13* and *6.14*.

Here, the feedback connection from IC_{1b} output to IC_{1a} input is made via

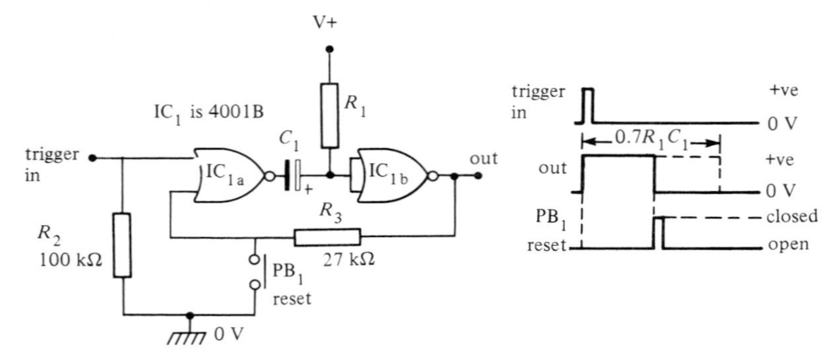

Figure 6.13 *Resettable NOR-type monostable*

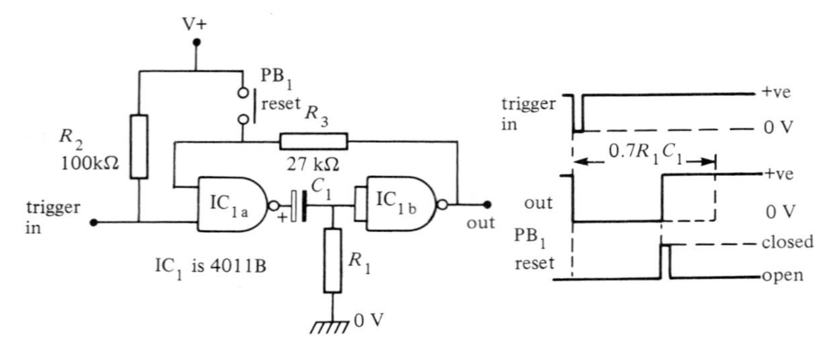

Figure 6.14 *Resettable NAND-type monostable*

R_3. Consequently, once the circuit has been triggered *and the original trigger signal has been removed* each circuit can be reset by forcing the feedback input of IC_{1a} to its normal quiescent state via push-button switch PB_1. In practice, PB_1 can easily be replaced by a transistor or CMOS switch etc., enabling the reset function to be accomplished via a suitable reset pulse.

Flip-flop monostables

Medium-accuracy monostables can easily be built by using standard edge-triggered CMOS flip-flop ICs such as the 4013B dual D-type or the 4027B dual JK-type in the configurations shown in *Figures 6.15* and *6.16*. Both of these circuits operate in the same basic way, with the IC wired in the frequency divider mode by suitable connection of its control terminals (data and set in the 4013B, and J, K, and set in the 4027B), but with the Q terminal connected

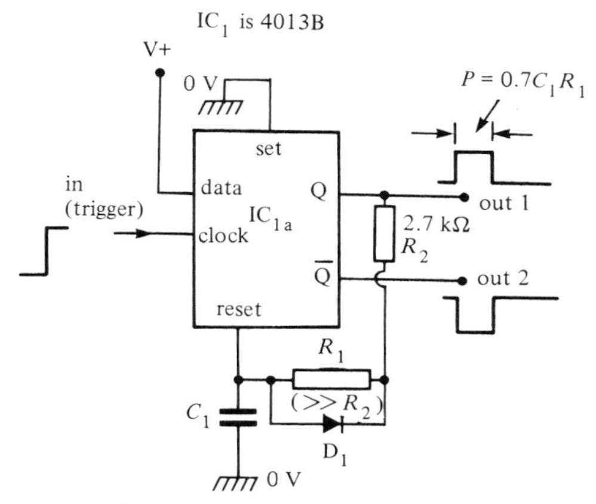

Figure 6.15 *D-type flip-flop used as a monostable*

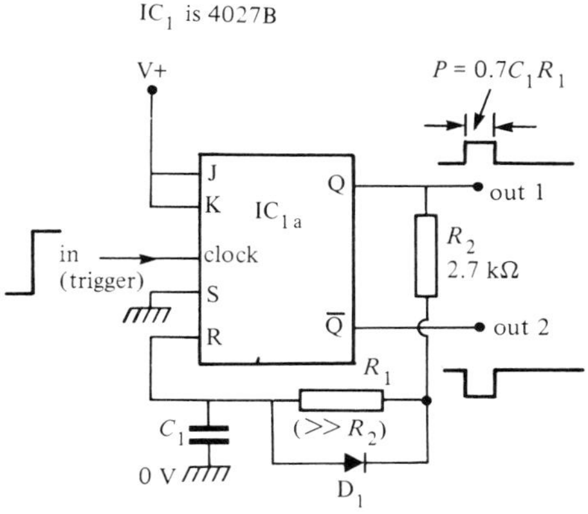

Figure 6.16 *JK-type flip-flop used as a monostable*

back to reset via a *C-R* time-delay network. The operating sequence of each circuit is as follows.

When the circuit is in its quiescent state the Q output terminal is low and discharges timing capacitor C_1 via R_2 and the parallel combination D_1-R_1. On the arrival of a sharply rising leading edge on the clock terminal the Q output flips high, and C_1 starts to charge up via the series combination R_1-R_2. Eventually, after a delay determined mainly by the C_1-R_2 values (R_1 is large relative to R_2), the C_1 voltage rises to such a value that the flip-flop is forced to reset, driving the Q terminal low again. C_1 then discharges rapidly via R_2 and D_1-R_1, and the circuit is then ready to generate another pulse on the arrival of the next trigger signal.

The timing period of the *Figure 6.15* and *6.16* circuits is roughly equal to 0.7 C_1R_1 and the reset period (the time taken for C_1 to discharge at the end of each pulse) roughly equals C_1R_2. In practice, R_2 is used mainly to prevent degradation of the trailing edge of the pulse waveform as C_1 discharges; R_2 can be reduced to zero if this degradation is acceptable. Note that the circuit generates a positive-going output pulse at Q, and a negative-going pulse at $\bar{Q}$ (not-Q), and the $\bar{Q}$ waveform is not influenced by the R_2 value.

The *Figure 6.15* and *6.16* circuits can be made resetable by connecting C_1 to the reset terminal via one input of an OR gate and using the other input of the OR gate to accept the external reset signal. *Figure 6.17* shows how the 4027B circuit can be so modified.

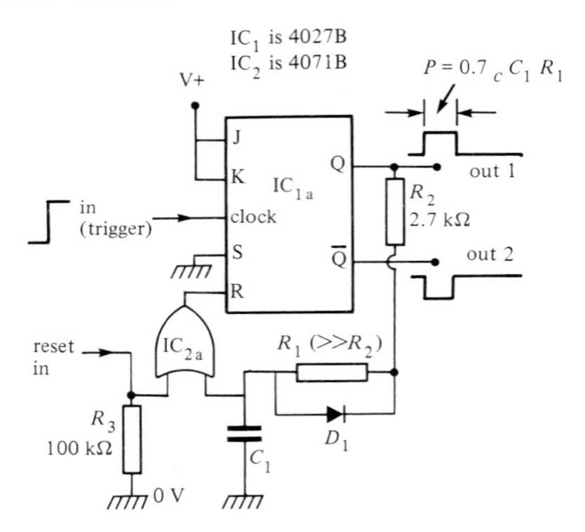

Figure 6.17 *Resettable JK-type monostable*

Finally, *Figure 6.18* shows how the 4027B can be used to make a retriggerable monostable in which the pulse period restarts each time a new trigger pulse arrives. Note that the input of this circuit is normally high, and that the circuit is actually triggered on the trailing (rising) edge of a negative-going input pulse. The circuit operates as follows.

At the start of each timing cycle the input trigger pulse switches low and rapidly discharges C_1 via D_1 and then, a short time later, the trigger pulse switches high again, releasing C_1 and simultaneously flipping the Q output high. The timing cycle then starts in the normal way, with C_1 charging via R_1 until the C_1 voltage rises to such a level that the flip-flop resets, driving the Q output low again and slowly discharging C_1 via R_1.

If a new trigger pulse arrives during a timing period (when Q is high and charging C_1 via R_1), C_1 discharges rapidly via D_1 on the low part of the

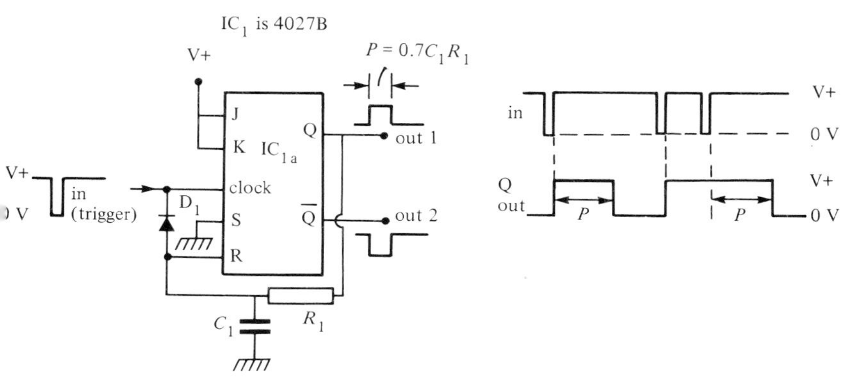

Figure 6.18 *Retriggerable JK-type monostable*

trigger, and commences a new timing cycle as the input waveform switches high again. In practice, the input trigger pulse must be wide enough to fully discharge C_1, but should be narrow relative to the width of the output pulse. The timing period of the output pulse equals $0.7C_1R_1$. For best results, R_1 should have as large a value as possible.

7555 monostables

In all the monostable circuits that we have looked at so far, the width of the output pulse depends on the threshold switching value of the IC that is used, and this value is subject to considerable variation between individual ICs and with variations in supply voltage and temperature. These circuits thus have only moderate accuracy. If very high pulse-width accuracy is needed, the best way of getting it is use a ICM7555 IC.

The ICM7555 is a CMOS version of the well known 555 timer chip, and uses a built-in supply-line-referenced precision voltage comparator to activate internal flip-flops and thus precisely control its triggered output pulse width, irrespective of wide variations in supply rail value and temperature. The 7555 can operate from supplies in the range 2 to 18 volts.

Figure 6.19 shows the basic way of using the 7555 as a manually triggered variable long-period pulse generator. Timing components R_1 C_1 are wired between the supply rails and have their junction taken to pins 6 and 7 of the IC. The IC is triggered by briefly pulling pin 2 low (to less than one-third of V_{supply}) via PB_1, at which moment the output (pin 3) switches high and the IC enters its timing cycle, with C_1 charging up via R_1. Eventually, after a delay of $1.1C_1R_1$, the C_1 voltage rises to the upper threshold switching value (two-thirds of V_{supply}) of the 7555, and the output switches abruptly low, ending the

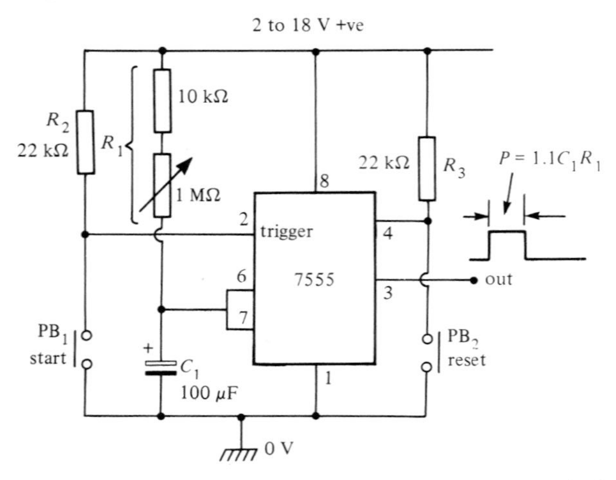

Figure 6.19 *Manually triggered 1.1 s to 100 s monostable with reset facility*

timing cycle. The timing cycle can be terminated prematurely, if required, by briefly pulling reset pin 4 low via PB_2.

In most practical applications of the 7555 the designer will want to trigger the IC electronically, rather than electromechanically, and in this case the trigger signal reaching pin 2 must be a carefully shaped negative-going pulse. Its amplitude must switch from an off value greater than two-thirds of V_{supply} to an on value less than one-third of V_{supply} (triggering actually occurs as pin 2 drops through the one-third of V_{supply} value). The pulse width must be greater

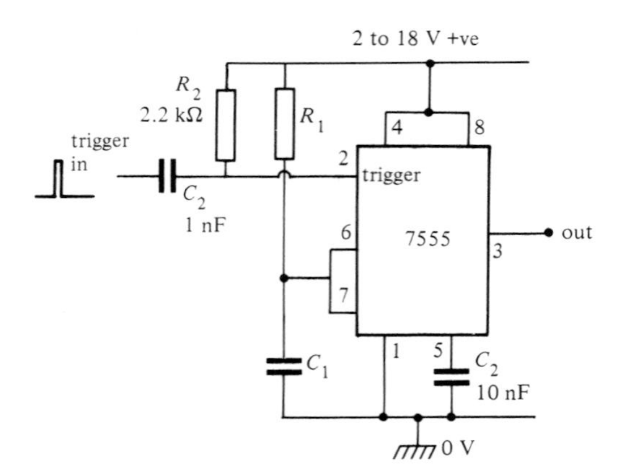

Figure 6.20 *Simple electronically triggered 7555 monostable*

than 100 ns but less than that of the desired output pulse, so that the trigger pulse is removed by the time the monostable period ends.

One way of generating a suitable trigger signal from a rectangular or square-wave input that switches fully between the supply rails is to connect the signal to pin 2 via a short-time-constant *C-R* differentiating network, which converts the leading or trailing edge of the waveform into suitable trigger pulses as shown in *Figure 6.20*.

The best possible way of triggering the 7555, however, is to use one of the previously described medium-accuracy monostables to generate a narrow (100 ns or greater) positive-going trigger pulse that is direct coupled to pin 2 of the 7555 via a transistor stage. *Figure 6.21* shows the connections. Note that in this and the previous circuit C_2 is used to decouple the trigger circuitry from the effects of supply line transients etc.

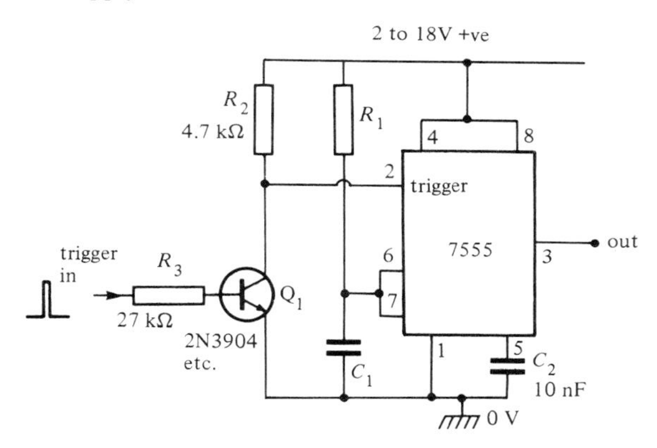

Figure 6.21 *Pulse-triggered 7555 monostable*

4047B and 4098B monostables

A number of dedicated CMOS monostable ICs are available and are worth considering in some circuit applications. The best known of these devices are the 4047B monostable/astable IC and the 4098B dual monostable (a greatly improved version of the 4528B). *Figure 6.22* shows the outlines and pin notations of these two devices.

Note that, like most of the CMOS-based monostables that we have already looked at, the 4047B and 4098B have rather poor pulse-width accuracy and stability. These ICs are, however, quite versatile, and can be triggered by either the positive or the negative edge of an input signal, and can be used in either the standard or the retriggerable mode.

The 4047B actually houses an astable multivibrator and a frequency divider stage, plus logic networks. When used in the monostable mode the

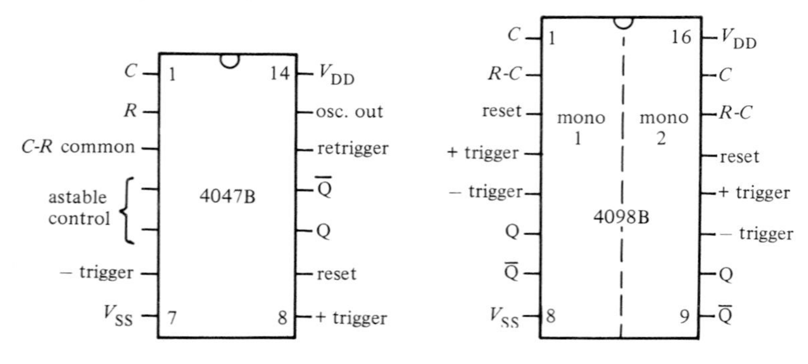

Figure 6.22 *Outlines and pin notations of the 4047B monostable/astable IC and the 4098B dual monostable*

trigger signal actually starts the astable and resets the counter, driving its Q output high. After a number of *C-R* controlled astable cycles the counter flips over and simultaneously kills the astable and switches the Q output low, completing the operating sequence. Consequently, the *C-R* timing components produce relatively long output pulse periods approximating 2.5*CR*.

In practice, *R* can have any value from 10 kΩ to 10 MΩ. *C* must be a non-polarized capacitor with a value greater than 1 nF. *Figures 6.23(a)* and *(b)* show how to connect the IC as a standard monostable triggered by either positive (a) or negative (b) input edges, and *Figure 6.23(c)* shows how to connect the monostable in the retriggerable mode. Note that these circuits can be reset at any time by pulling pin 9 high.

The 4098B is a fairly simple dual monostable, in which the two mono sections share common supply connections but can otherwise be used independently. Mono 1 is housed on the left side (pins 1 to 7) of the IC and mono 2 on the right side (pins 9 to 15) of the IC. The timing period of each mono is controlled by a single resistor *R* and capacitor *C*, and approximates 0.5*CR*. *R* can have any value from 5 kΩ to 10 MΩ, and *C* can have any value from 20 pF to 100 μF. *Figure 6.24* shows a variety of ways of using the 4098B. Note that in these diagrams the bracketed numbers relate to the pin connections of mono 2 and the plain numbers to mono 1, and that the reset terminal (pins 3 or 13) is shown disabled.

Figures 6.24(a) and *(b)* show how to use the IC to make retriggerable monostables that are triggered by positive or negative input edges, respectively. In *Figure 6.24(a)* the trigger signal is fed to the +trigger pin and the −trigger pin is tied low. In *Figure 6.24(b)* the trigger signal is applied to the −trigger pin and the +trigger pin is tied high.

Figures 6.24(c) and *(d)* show how to use the IC to make standard (non-retriggerable) monostables that are triggered by positive or negative edges, repectively. These circuits are similar to those mentioned above except that

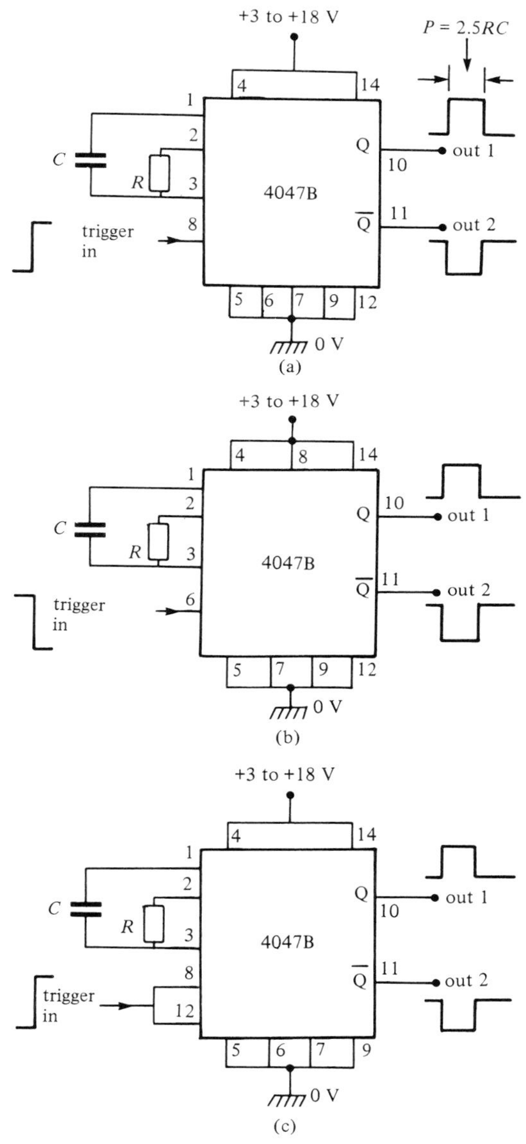

Figure 6.23 *Various ways of using the 4047B as a monostable: (a) positive-edge-triggered monostable (b) negative-edge-triggered monostable (c) positive-edge triggered monostable, retriggerable*

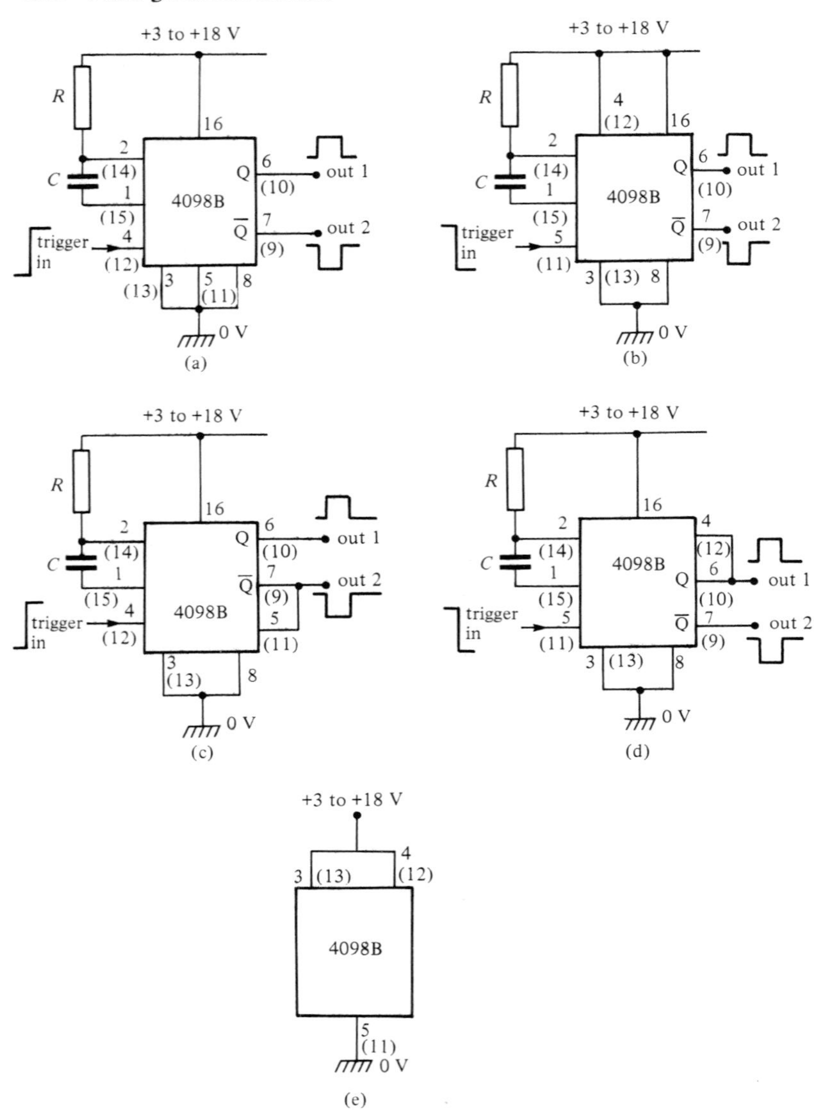

Figure 6.24 *Various ways of using the 4098B monostable: (a) positive-edge-triggering mono, retriggerable (b) negative-edge-triggering mono, retriggerable (c) positive-edge-triggering mono, non-retriggerable (d) negative-edge-triggering mono, non-retriggerable (e) connections for each unused section of the IC*

the unused trigger pin is coupled to either the Q or the not-Q output, so that trigger pulses are blocked once a timing cycle has been initiated.

Finally, *Figure 6.24(e)* shows how the unused half of the IC must be connected when only a single monostable is wanted from the package. The −trigger pin is tied low, and the +trigger and reset pins are tied high.

7 Clocked flip-flops

CMOS digital ICs can be classified into two basic types: those based on simple gate networks (logic types), and those based primarily on clocked flip-flop (bistable or memory) elements. The latter include simple counter/dividers, shift registers and data latches, and complex devices such as presettable up/down counters and dividers; in all these cases, however, the devices are based on the simple clocked flip-flop element.

In this chapter we explain how the clocked flip-flop circuit works, and then go on to introduce some simple CMOS flip-flop ICs and show some practical ways of using them. As an immediate follow-up the next two chapters introduce a range of advanced counter/divider ICs and associated devices, together with a stack of applications information.

Basic principles

The simplest type of CMOS flip-flop is the cross-coupled bistable. *Figure 7.1* shows the circuit, symbol and truth table of a NOR gate version of this

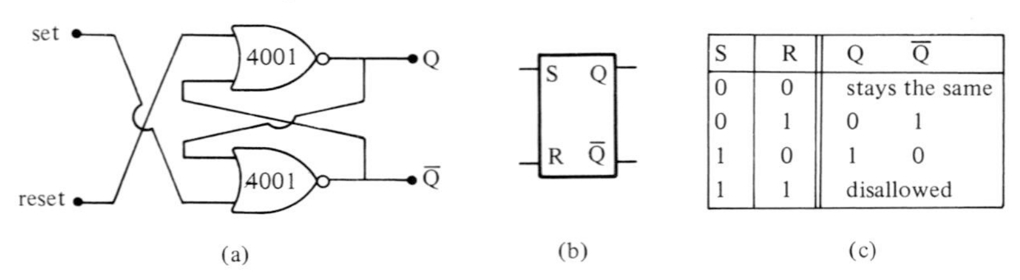

S	R	Q	$\overline{Q}$
0	0	stays the same	
0	1	0	1
1	0	1	0
1	1	disallowed	

Figure 7.1 *(a) Circuit, (b) symbol and (c) truth table of a NOR-type (S-R) flip-flop*

device, which has two input terminals (normally tied low via pull-down resistors) and a pair of anti-phase output terminals (Q and not-Q or $\bar{Q}$).

The basic action of the circuit is such that if the set terminal is briefly taken high (to logic-1) the Q output immediately switches high (and the not-Q output switches low), and the cross-coupling then causes the outputs to latch into this state even when both inputs are pulled low again. The only way that the output states can be changed is to apply a logic-1 to the reset terminal, in which case the Q output immediately switches low (and the not-Q output switches high), and the cross-coupling then causes the outputs to latch into this new state even when both inputs are pulled low again.

Thus, the basic set-reset (S-R) flip-flop acts as a simple memory element which 'remembers' which of the two inputs last went high. Note that if both inputs go high simultaneously the output states cannot be predicted, so this event must not be allowed to occur.

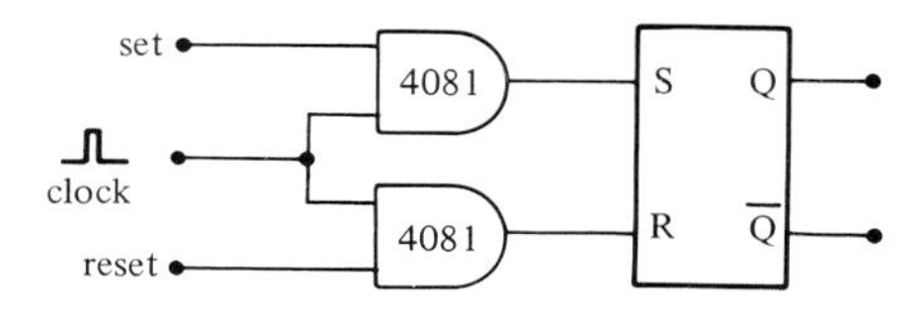

Figure 7.2 *Clocked S-R flip-flop*

The versatility of the basic *Figure 7.1* circuit can be greatly enhanced by wiring an AND gate in series with each input terminal, using the connections shown in *Figure 7.2*, so that high input signals can only reach the S-R flip-flop when the clock signal is also high. Thus, when the clock signal is low, both inputs of the S-R flip-flop are held low, irrespective of the states of the set and reset inputs, and the flip-flop acts as a permanent memory, but when the clock signal is high the circuit acts as a standard R-S flip-flop. Consequently, information is not automatically latched into the flip-flop, but must be clocked in via the clock terminal: this circuit is thus known as a clocked S-R flip-flop.

Figure 7.3 shows how two of these clocked S-R flip-flops can be cascaded and clocked in anti-phase (via an inverter in the clock line) to make the most important of all flip-flop elements, the so-called clocked master-slave flip-flop. The basic action of this circuit is as follows.

Master-slave flip-flop

When the clock input terminal of *Figure 7.3* is in the low state the inputs to the master flip-flop are enabled via the inverter, so the set-reset data is accepted, but the inputs to the slave flip-flop are disabled, so this data is not passed to

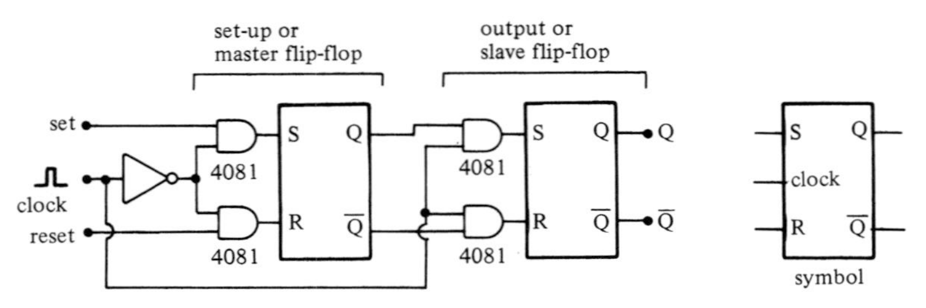

Figure 7.3 *Clocked master-slave flip-flop circuit and symbol*

the output terminals. When the clock input terminal goes to the high state the inputs to the master flip-flop are disabled via the inverter, which thus outputs only the remembered input data, and simultaneously the input to the slave flip-flop is enabled and the remembered data is latched and passed to the output terminals.

Thus, the clocked master-slave flip-flop accepts input data or information only when the clock signal is low, and passes that data to the output on the arrival of the leading edge of a positive-going clock signal (edge clocking). The clocked master-slave flip-flop is such an important device that it is given its own circuit symbol, as shown.

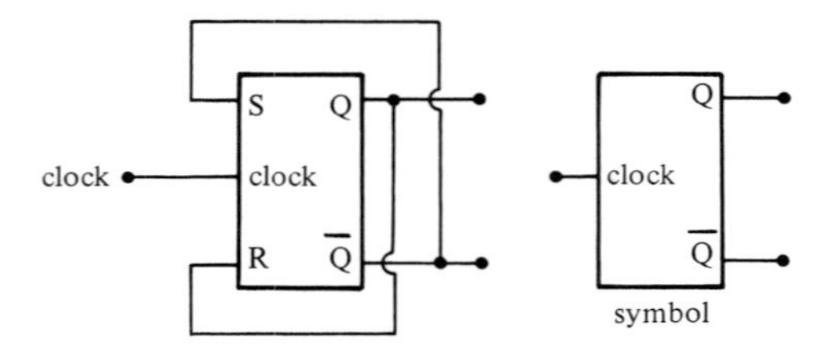

Figure 7.4 *A toggle or type-T flip-flop is a* Figure 7.3 *type configured as shown*

The clocked master-slave flip-flop can be made to give a toggle (divide-by-two) action by cross-coupling input and output terminals as shown in *Figure 7.4*, so that set and Q (and reset and not-Q) logic levels are always opposite. Consequently, when the clock signal is low the master flip-flop receives the instruction 'change state', and when the clock goes high the slave flip-flop executes the instruction; so the output changes state on the arrival of the leading edge of each new clock pulse. It takes two clock pulses to change the output from one state to another and then back again, so the output switching frequency is half that of the clock frequency. This circuit, which is known as a toggle or type-T flip-flop, thus acts as a binary counter/divider.

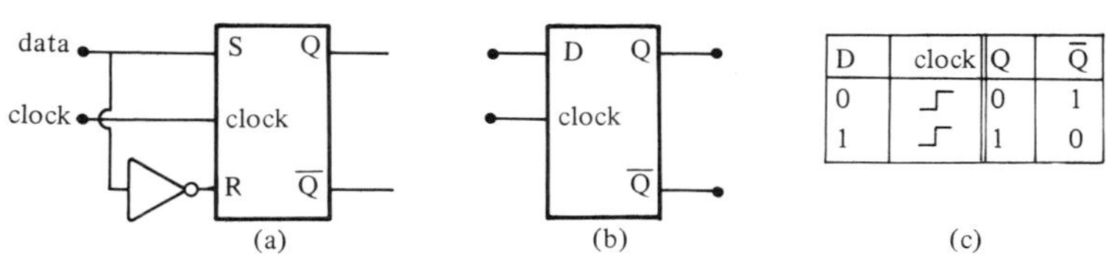

D	clock	Q	$\overline{Q}$
0	↑	0	1
1	↑	1	0

Figure 7.5 *(a) Basic circuit, (b) symbol and (c) truth table of a simple D-type flip-flop*

D and JK flip-flops

The type-T flip-flop is a specialized element which acts purely as a counter/divider. A far more versatile device is the data or type-D flip-flop, which is made by connecting the clocked master-slave flip-flop in the configuration shown in *Figure 7.5*. Here an inverter is wired between the S and R terminals of the flip-flop, so that these terminals are always in anti-phase and the input data is applied via a single data pin. *Figures 7.5(b)* and *(c)* show the symbol and truth table of the type-D flip-flop. It can be used as a data latch by using the connections shown in *Figure 7.6(a)*, or as a binary counter/divider by using the connections shown in *Figure 7.6(b)* (with the D and not-Q terminals coupled together).

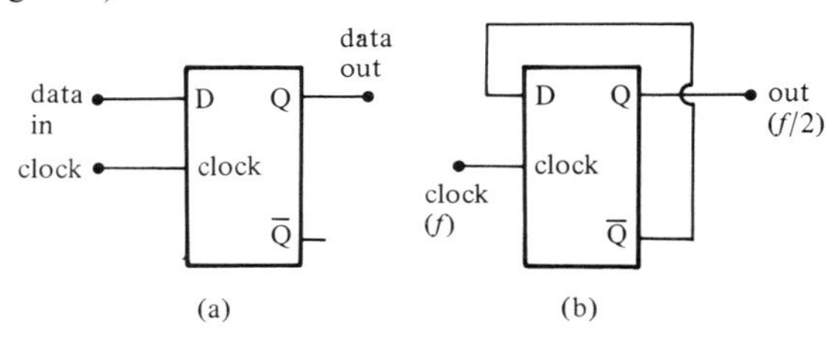

Figure 7.6 *A D-type flip-flop can be used as (a) a data latch or (b) a divide-by-two (binary counter/divider) circuit*

Figure 7.7 shows the basic circuit, symbol and action table of an even more important and versatile clocked flip-flop, which is universally known as the JK-type flip-flop. This flip-flop can be 'programmed' to act as either a data latch, a counter/divider, or a do-nothing element by suitably connecting the J and K terminals as indicated in the table. In essence, the JK flip-flop acts like a T-type when both J and K terminals are high, or as a D-type when the J and K terminals are at different logic levels. When both J and K terminals are low the flip-flop states remain unchanged on the arrival of a clock pulse.

4013B and 4027B flip-flops

The two best known CMOS clocked flip-flop ICs are the 4013B D-type and the 4027B JK-type. Both of these ICs are duals, each containing two

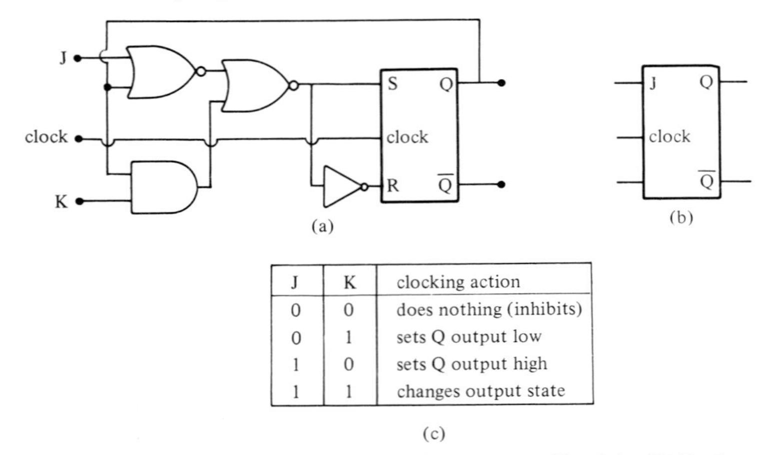

J	K	clocking action
0	0	does nothing (inhibits)
0	1	sets Q output low
1	0	sets Q output high
1	1	changes output state

(c)

Figure 7.7 *(a) Basic circuit, (b) symbol and (c) action table of the JK flip-flop*

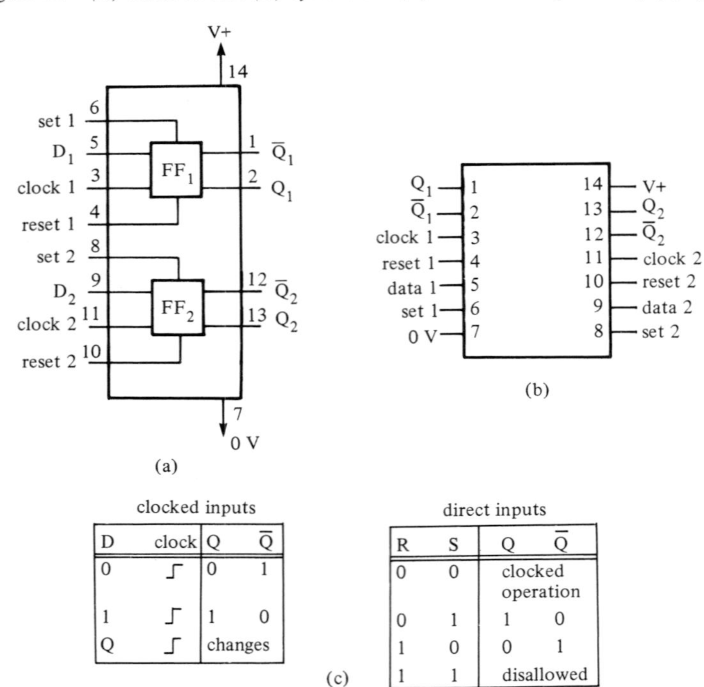

clocked inputs

D	clock	Q	$\overline{Q}$
0	⌐	0	1
1	⌐	1	0
Q	⌐	changes	

direct inputs

R	S	Q	$\overline{Q}$
0	0	clocked operation	
0	1	1	0
1	0	0	1
1	1	disallowed	

(c)

Figure 7.8 *(a) Functional diagram, (b) outline and (c) truth tables of the 4013B dual D-type flip-flop*

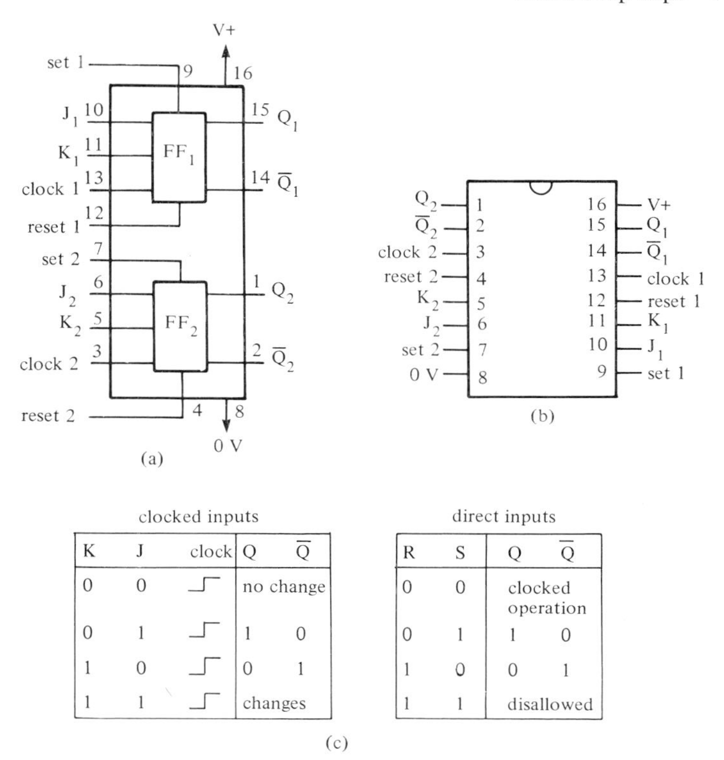

clocked inputs

K	J	clock	Q	$\overline{Q}$
0	0	⎍	no change	
0	1	⎍	1	0
1	0	⎍	0	1
1	1	⎍	changes	

direct inputs

R	S	Q	$\overline{Q}$
0	0	clocked operation	
0	1	1	0
1	0	0	1
1	1	disallowed	

(c)

Figure 7.9 *(a) Functional diagram, (b) outline and (c) truth tables of the 4027B dual JK flip-flop*

independent flip-flops sharing common power supply connections. *Figure 7.8* shows the functional diagram, pin connections, and truth tables of the 4013B, and *Figure 7.9* shows similar details of the 4027B.

Note that both of these ICs have set and reset input terminals that are additional to the connections shown in the basic *Figures 7.5* and *7.7* circuits. These terminals are known as direct inputs and enable the clock action of the flip-flop to be overridden, so that the devices can act as simple unclocked set-reset flip-flops. For normal clocked operation (as a counter/divider or data latch etc.), the direct R and S terminals must be tied to logic-0 as indicated.

The 4013B and 4023B are fast-acting ICs, and when using them it is very important to note that their clock signals must be absolutely clean (noise-free and bounceless) and have rise and fall times of less than 5 μs. The 4013B is particularly fussy about the shape of its input clock signals, the 4027B rather less so. Both devices clock or shift on the positive transition of the clock signal.

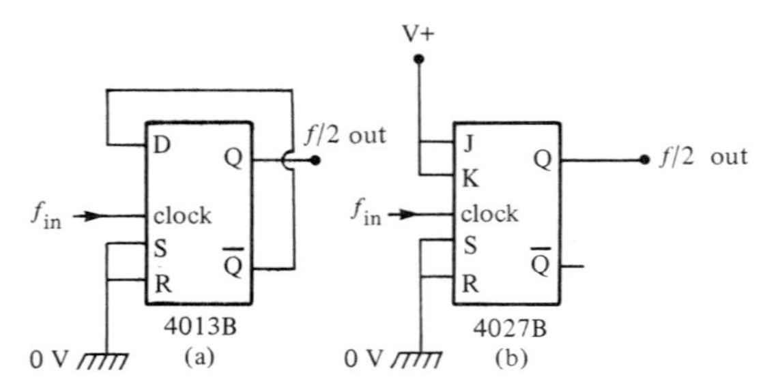

Figure 7.10 *Divide-by-two counters made from (a) D and (b) JK flip-flop stages*

Ripple counters

The most popular application of the clocked flip-flop is as a binary counter, and *Figure 7.10* shows how to connect the 4013B D-type and 4027B JK-type flip-flops to make divide-by-two counters. In both cases the S and R terminals are tied to logic-0 to accept clocked operation. In the case of the 4013B the not-Q output is tied directly to the data input terminal, while in the case of the 4027B the J and K terminals are both tied high (to logic-1) to give counter/divider action. When clocked by a fixed-frequency waveform both circuits give a symmetrical square-wave output at half of the clock frequency.

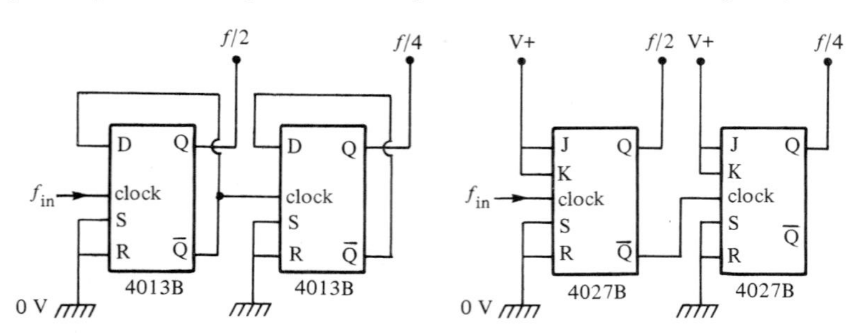

Figure 7.11 *D and JK versions of divide-by-four ripple counters*

Numbers of the *Figure 7.10* stages can be cascaded to give multiple binary division by simply clocking each new stage from the not-Q output of the preceding stage. Thus *Figure 7.11* shows how two D or JK stages can be cascaded to give an overall division ratio of four (2^2), and *Figure 7.12* shows how three stages can be cascaded to give a division ratio of eight (2^3). *Figure 7.13* shows how D stages can be cascaded to make a divide-by-2^N counter, where N is the number of counter stages. Thus, four stages give a ratio of sixteen (2^4), five stages give 32 (2^5), six give 64 (2^6), and so on.

The *Figure 7.11* to *7.13* circuits are known as ripple counters, because each

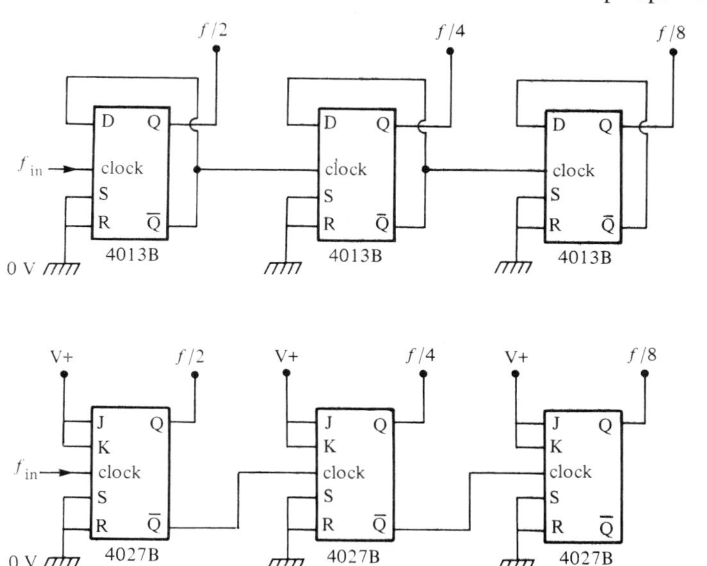

Figure 7.12 *D and JK versions of divide-by-eight counters*

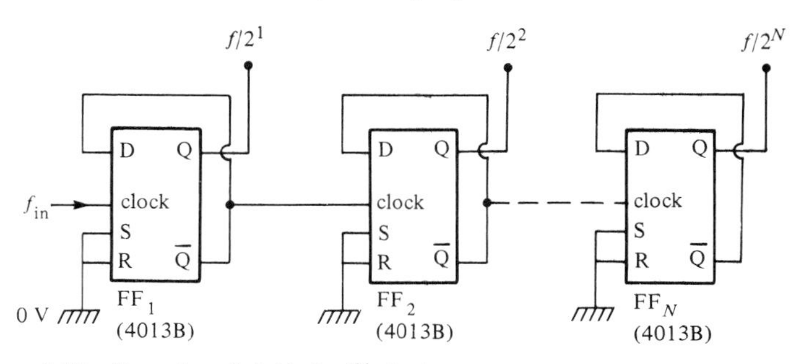

Figure 7.13 *D version of divide-by-2ᴺ ripple counter*

stage is clocked by the preceding stage (rather than directly by the input clock signal), and the clock signal thus seems to ripple through the counter. Inevitably, the propagation delays of the individual dividers all add together to give a summed delay at the end of the chain, and counter stages (other than the first) thus do not clock in precise synchrony with the original clock signal; such counters are thus asynchronous in action. If the outputs of the counter stages are decoded via gate networks, the propagation delays of the asynchronous counter can result in unwanted output signals (see section 'Decoding' in this chapter).

Long ripple counters

Type 4013B and 4027B counters can be cascaded to give any desired number of ripple stages. Where more than four stages are needed, however, it is usually economic to use a special-purpose MSI ripple-carry binary counter/divider IC. *Figures 7.14* to *7.17* show the outlines and functional diagrams of four popular ICs of this type.

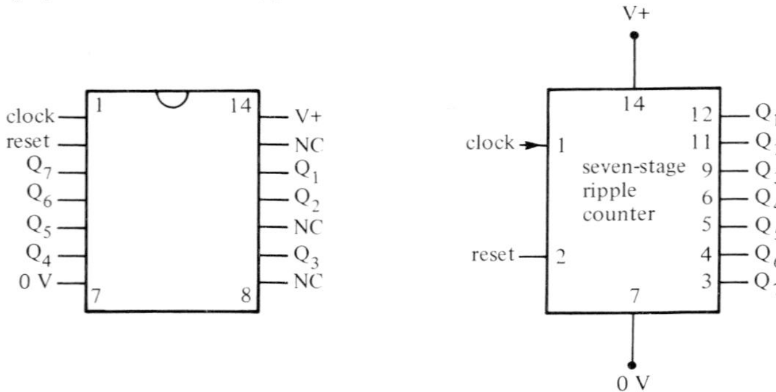

Figure 7.14 *Outline and functional diagram of the 4024B seven-stage ripple counter*

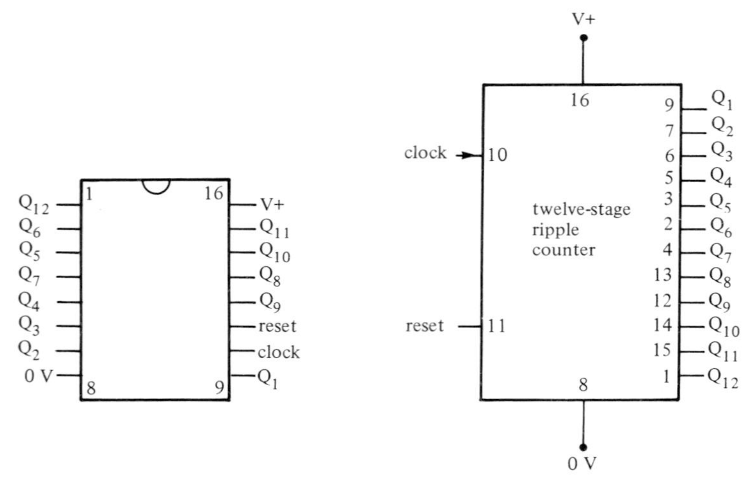

Figure 7.15 *Outline and functional diagram of the 4040B twelve-stage ripple counter*

The 4024B (*Figure 7.14*) is a seven-stage ripple unit with all seven outputs externally accessible; it gives a maximum division ratio of 128. The 4040B (*Figure 7.15*) is a twelve-stage unit with all outputs accessible; it gives a maximum division ratio of 4096. The 4020B (*Figure 7.16*) is a fourteen-stage

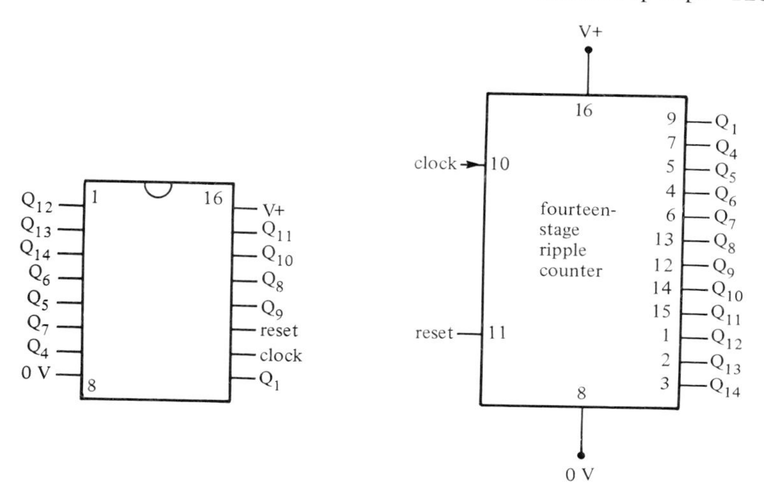

Figure 7.16 *Outline and functional diagram of the 4020B fourteen-stage ripple counter*

unit with all outputs except 2 and 3 externally accessible; it gives a maximum division ratio of 16384.

Figure 7.17 gives details of the 4060B. This is another fourteen-stage unit, but does not have outputs 1, 2, 3 and 11 externally accessible. A special feature of this IC is that it incorporates a built-in clock oscillator circuit. The diagram shows the connections for using the internal circuit as either a crystal or an *R-C* oscillator.

Note that the 4020B, 4024B, 4040B, and 4060B ripple counters are all provided with Schmitt trigger action on their input terminals, and can thus be clocked via slow or non-rectangular input waveforms. They all trigger on the negative transition of each input pulse. All counters can be set to zero by applying a high level (logic-1) to the reset line.

Decoding

The outputs of a two-stage divide-by-four ripple counter (*Figure 7.18(a)*) have four possible coded states, as shown in *Figure 7.18(b)*. Thus, at the start or 0 reference point of each clock cycle the Q_2 and Q_1 outputs are both in the logic-0 state. On the arrival of the first clock pulse in the cycle, Q_1 switches high. On the arrival of the second pulse, Q_2 goes high and Q_1 goes low. On the third pulse, Q_2 and Q_1 both go high. Finally on the arrival of the fourth pulse Q_2 and Q_1 both go low again, and the cycle is back to its original 0 reference state.

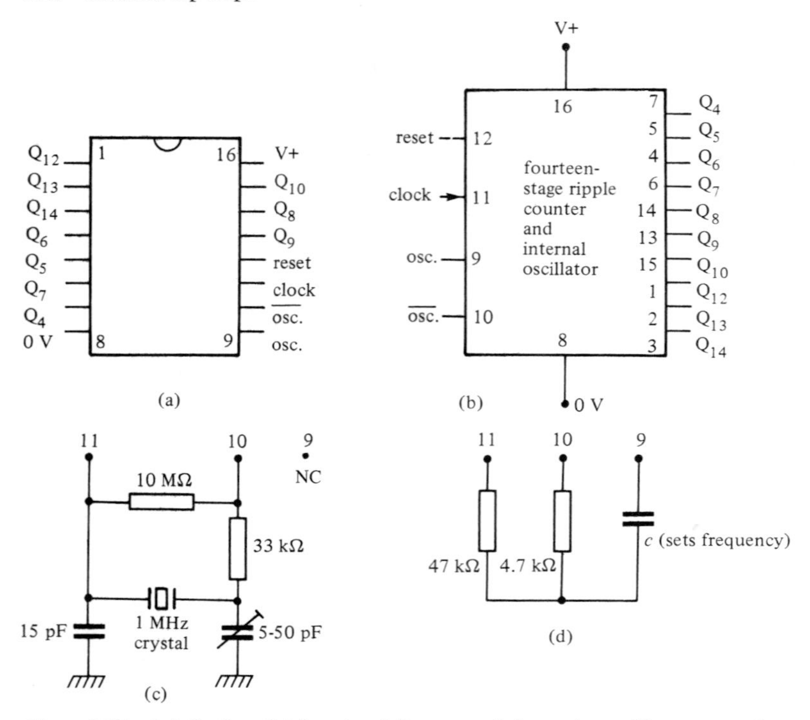

Figure 7.17 *(a) Outline, (b) functional diagram and alternative oscillator connections of the 4060B fourteen-stage ripple counter (c) crystal oscillator (d)* R-C *oscillator*

Each of the four possible coded states of the ripple counter can be decoded, to give four unique outputs, by ANDing the high outputs that are unique to each state, as shown in *Figure 7.18(c)*. Since the ripple counter is an asynchronous device, however, the propagation delay between the two flip-flops may cause signals or 'glitches' to appear in the decoded outputs, as illustrated by the 0 decoded waveforms of *Figure 7.18(d)*.

The principles outlined in *Figure 7.18* can be extended to any multistage ripple counter in which the coded outputs are accessible for decoding. Note, however, that the greater the number of stages the greater the total propagation delays and, consequently, the greater the magnitude of the decoded glitches.

Up and down counters

A standard ripple counter is one in which positive-edge-clocked flip-flops are used, and in which each stage (except the first) is clocked from the not-Q

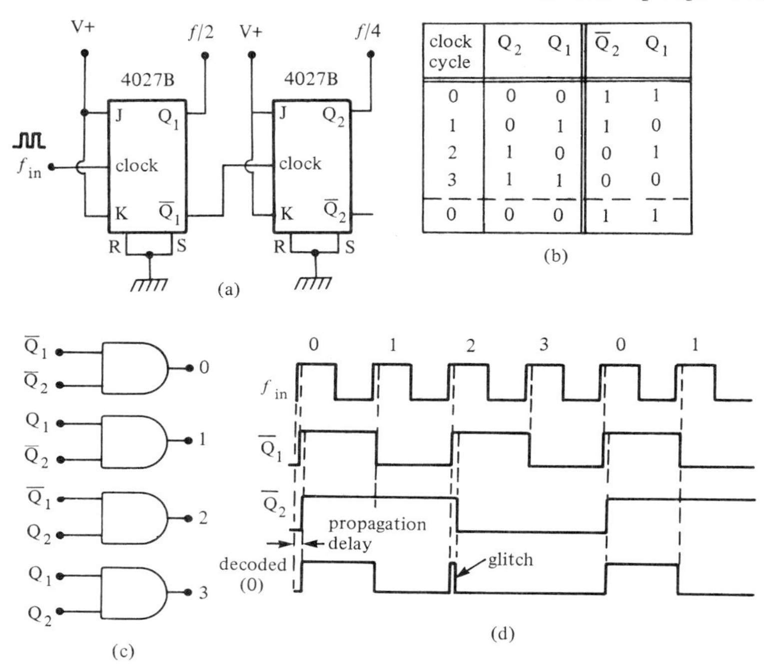

clock cycle	Q_2	Q_1	$\overline{Q}_2$	Q_1
0	0	0	1	1
1	0	1	1	0
2	1	0	0	1
3	1	1	0	0
0	0	0	1	1

Figure 7.18 *(a) Circuit and (b) coded output states of a two-stage ripple counter. Each of the four possible coded states can be decoded via a two-input AND gate (c), but in a ripple counter the decoded outputs may not be glitch-free (d)*

output of the preceding stage. As shown from the coded output states of *Figure 7.18(b)* the binary outputs of a standard counter increase with each succeeding pulse of the clock cycle, and such counters are inherently known as up or add counters.

It is, however, possible to make down or subtract counters, in which the binary-coded output decreases with each new clock pulse, by simply clocking each flip-flop stage (except the first) from the Q output of the preceding stage. *Figure 7.19* shows the circuit and truth table of a two-stage (divide-by-four) ripple down or subtract counter.

Walking-ring (Johnson) counters

Ripple counters are very useful where undecoded binary division is needed, but (because of the glitch problems) are not very suitable for use in decoded counting applications. Fortunately, an alternative dividing technique that is suitable for use in decoded counting applications is available. It is known

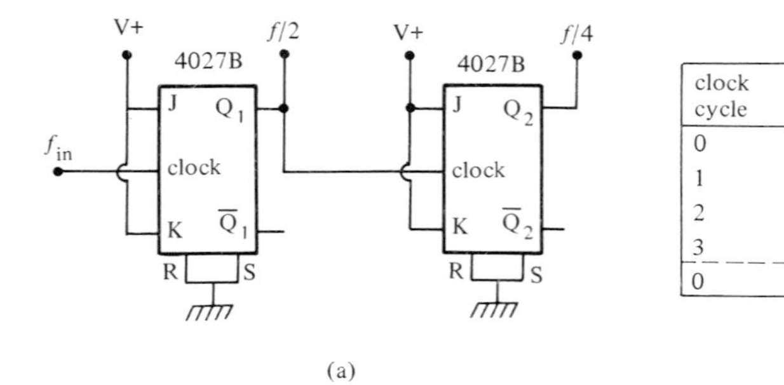

clock cycle	Q_2	Q_1
0	0	0
1	1	1
2	1	0
3	0	1
0	0	0

(a) (b)

Figure 7.19 *(a) Circuit and (b) truth table of a two-stage (divide-by-four) ripple down counter*

as the walking ring or Johnson technique. Such circuits rely on the programmable nature of JK flip-flops that enables them to act as either set or reset latches, as binary dividers, or as do-nothing devices. In walking-ring (Johnson) counters all flip-flops are clocked in parallel, and thus operate in synchrony with the input clock signal, and are known as synchronous counters. They give glitch-free outputs.

Figure 7.20 shows the circuit and truth tables of a synchronous divide-by-three counter. Note that the truth table shows the action state of each flip-flop at each stage of the counting cycle; remember that when the clock is low the action instruction is loaded (via the JK terminals) into the flip-flop, and the instruction is then carried out as the clock signal transitions high.

Thus at the start of the cycle (clock low), when Q_2 and Q_1 are both low the instruction 'change state' (11) is loaded into FF_1, and the instruction 'set Q_2 low' (01) is loaded into FF_2. On the arrival of the first clock pulse this instruction is carried out, and Q_1 goes high and Q_2 stays low.

When the clock goes low again, new program information is fed to the flip-flops. FF_1 is instructed to 'change state' (11) and FF_2 is instructed 'set Q_2 high' (10); these instructions are implemented on the positive transition of the second clock pulse, causing Q_2 to go high and Q_1 to go low. When the clock goes low again new program information is again fed to the flip-flops from the outputs of their partners. FF_1 is instructed 'set Q_1 low' (01) and FF_2 is instructed 'set Q_2 low' (01); these instructions are implemented on the positive transition of the next clock pulse, causing Q_1 and Q_2 to go back to their original 0 states. The counting sequence then repeats *ad infinitum*.

Thus in the walking-ring (Johnson) counter all flip-flops are clocked in parallel, but are cross-coupled so that the clocking response of any one stage depends on the states of the other stages. Walking-ring counters can be configured to give any desired count ratio, and *Figures 7.21* and *7.22* show the circuits and truth tables of divide-by-four and divide-by-five counters respectively.

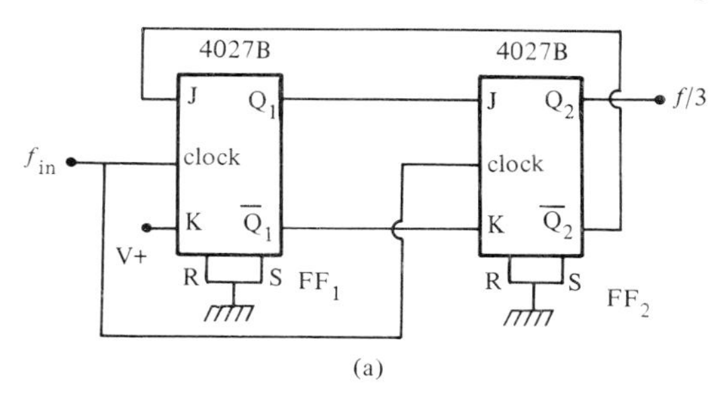

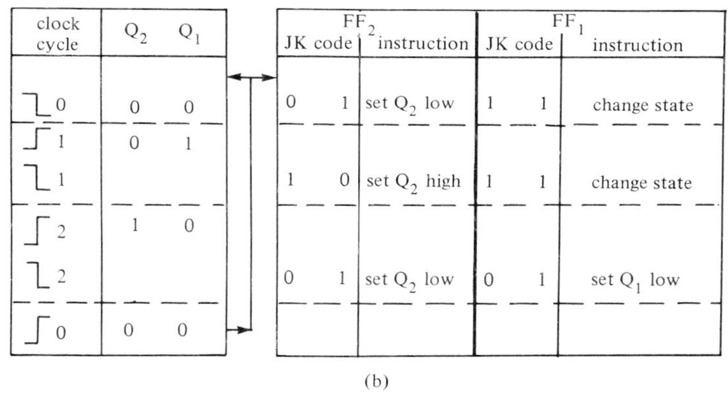

clock cycle	Q_2	Q_1	FF_2 JK code		FF_2 instruction	FF_1 JK code		FF_1 instruction
↓ 0	0	0	0	1	set Q_2 low	1	1	change state
↑ 1	0	1						
↓ 1			1	0	set Q_2 high	1	1	change state
↑ 2	1	0						
↓ 2			0	1	set Q_2 low	0	1	set Q_1 low
↑ 0	0	0						

(b)

Figure 7.20 *(a) Circuit and (b) truth tables of a synchronous divide-by-three counter*

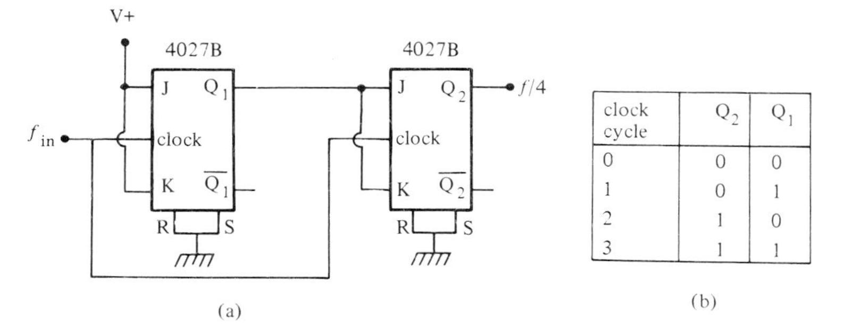

clock cycle	Q_2	Q_1
0	0	0
1	0	1
2	1	0
3	1	1

(b)

Figure 7.21 *(a) Circuit and (b) truth table of a synchronous divide-by-four counter*

4018B divide-by-*N* counter

When synchronous count numbers greater than four are needed it is usually economical to use an MSI CMOS IC such as the 4018B (rather than several

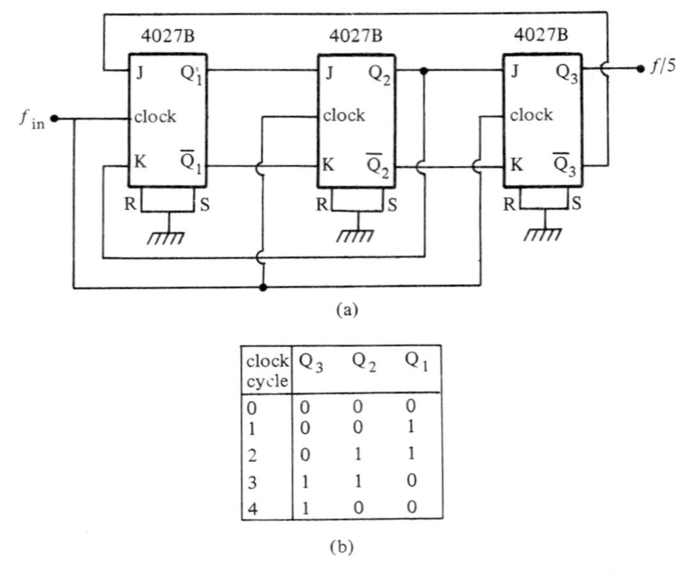

clock cycle	Q_3	Q_2	Q_1
0	0	0	0
1	0	0	1
2	0	1	1
3	1	1	0
4	1	0	0

(b)

Figure 7.22 *(a) Circuit and (b) truth table of synchronous divide-by-five counter*

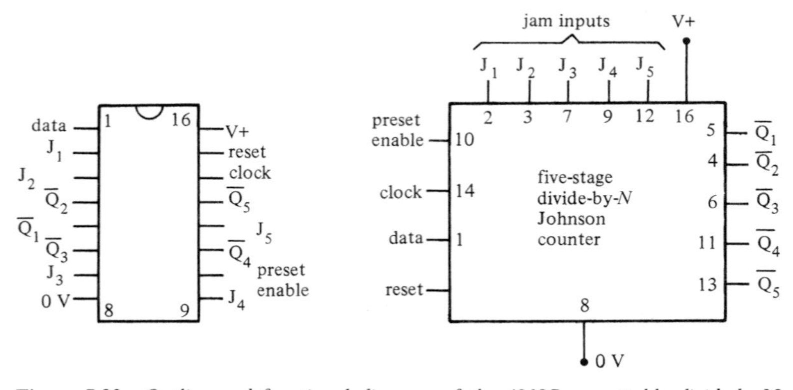

Figure 7.23 *Outline and functional diagram of the 4018B presettable divide-by-*N counter

4027Bs) to perform the function. *Figure 7.23* shows the functional diagram and outline of the 4018B presettable divide-by-N counter; it can be made to divide by any whole number between two and ten by merely cross-coupling its data and output terminals in various ways.

The 4018B incorporates a five-stage Johnson counter, has a built-in Schmitt trigger in its clock line, and clocks on the positive transition of the

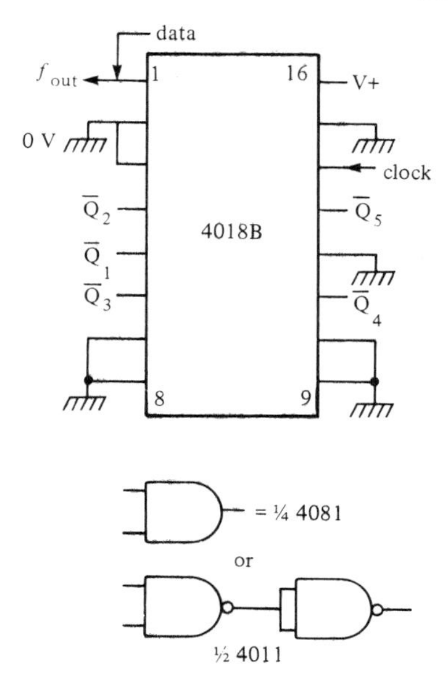

division ratio	feedback connections
2	$\overline{Q}_1$ to data
3	$\overline{Q}_1$, $\overline{Q}_2$ → AND gate → to data
4	$\overline{Q}_2$ to data
5	$\overline{Q}_2$, $\overline{Q}_3$ → AND gate → to data
6	$\overline{Q}_3$ to data
7	$\overline{Q}_3$, $\overline{Q}_4$ → AND gate → to data
8	$\overline{Q}_4$ to data
9	$\overline{Q}_4$, $\overline{Q}_5$ → AND gate → to data
10	$\overline{Q}_5$ to data

Figure 7.24 *Methods of connecting the 4018B for divide-by-two to divide-by-ten operation*

input signal. The counter is said to be presettable because its outputs can be set to a desired state at any time by feeding the inverted version of the desired binary code to the J_1 to J_5 'Jam' inputs and then loading the data by taking the preset enable (pin 10) terminal high.

Figure 7.24 shows methods of connecting the 4018B to give any whole-number division ratio between two and ten. On even division ratios no additional components are needed, but on odd ratios a two-input AND gate is needed in the feedback network; this can be a single 4081B AND gate, or can be made from two 4011B NAND gates.

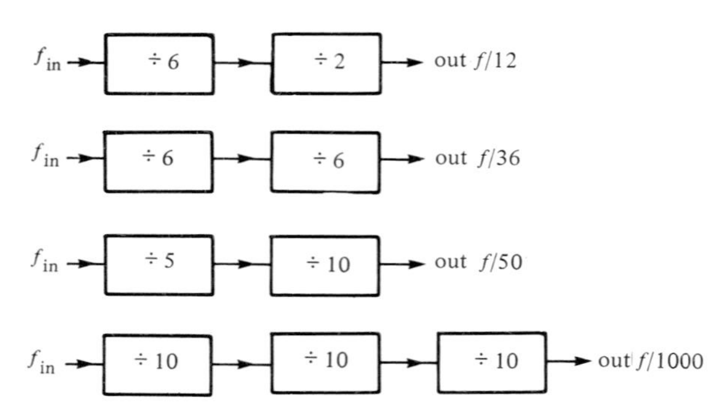

Figure 7.25 *Typical examples of division by numbers greater than ten*

Greater-than-ten division

Even division ratios greater than ten can usually be obtained by simply cascading suitably scaled counter stages, as shown in *Figure 7.25*. Thus, a divide-by-two and a divide-by-six stage give a ratio of twelve, a divide-by-six and a divide-by-six give a ratio of 36, and so on. Non-standard and uneven division ratios can be obtained by using standard synchronous counters (such as the 4018B) and decoding their outputs to generate suitable counter-reset pulses on completion of the desired count. More advanced types of counter, together with special decoder ICs, are discussed in the next two chapters.

Latches and registers

To round off this chapter, let's move away from counters and take a brief look at three other applications of the clocked master-slave flip-flop. *Figure 7.26* shows how to make a four-bit data latch from four D-type flip-flop stages (for an eight-bit latch, use eight flip-flops). The data latch is useful for storing numbers, data, etc. Input data is ignored until a positive store pulse is applied,

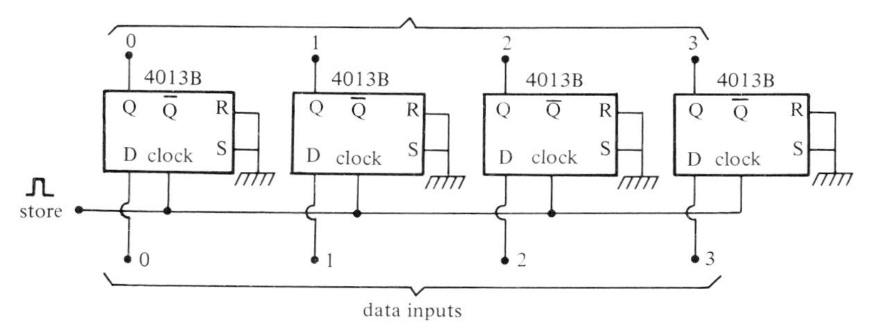

Figure 7.26 *Four-bit data latch*

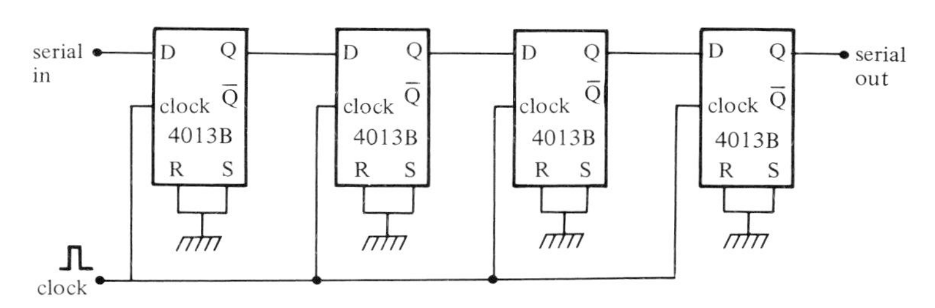

Figure 7.27 *Four-bit serial-in/serial-out (SISO) shift register*

at which point the latch stores and outputs that data and maintains it until told to do otherwise via a new store command.

Figure 7.27 shows how to make a four-bit serial-in/serial-out (SISO) shift register. If a bit of binary data is applied to the input it is passed to the output of the first flip-flop on the application of the first clock pulse, then to the output of the second on the second pulse, to the output of the third on the

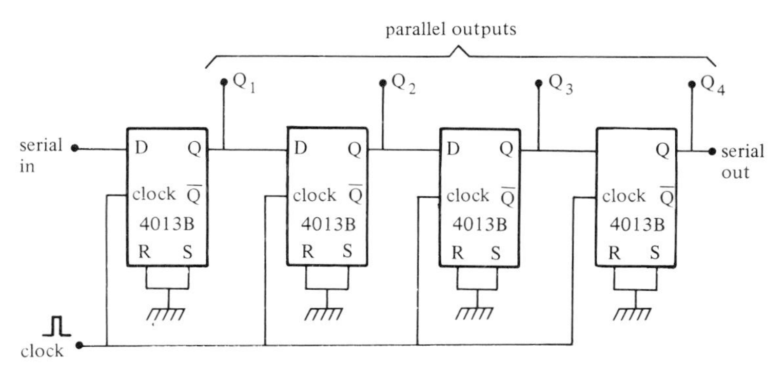

Figure 7.28 *Four-bit serial-in/parallel-out (SIPO) shift register*

third pulse, and to the fourth (final) output on the fourth pulse. The circuit can hold four bits of data at any given moment. The SISO register is useful for simply delaying binary signals, or for storing bits of binary data and unloading them (in serial form) when required.

Finally, *Figure 7.28* shows how the above circuit can be converted to a serial-in/parallel-out (SIPO) shift register by simply taking the parallel outputs from the Q outputs of all flip-flops. This register is useful for converting serial data into parallel form.

8 Up and up/down counters

In the previous chapter we looked in detail at CMOS D and JK flip-flops and showed how they can be used to make a variety of types of simple counting/dividing and other circuits. In this chapter we continue this theme by looking at sophisticated CMOS up and up/down counter/dividers. We round off the subject in Chapter 9 by looking at dedicated down counters and at various types of decoders and display drivers.

The 4017B decade counter

The best known sophisticated CMOS up counter is the 4017B. This is a five-stage synchronous walking-ring or Johnson decade counter with ten fully decoded outputs that switch high sequentially on the arrival of each new clock pulse, only one output being high at any moment. Each output can sink or source up to several milliamperes of current.

Figure 8.1 shows the outline and functional diagram of the 4017B, and *Figure 8.2* shows the basic timing diagram of the device, which has clock, reset, and inhibit input terminals and ten decoded and one carry output terminal.

The 4017B's counters are advanced one step at each positive transition of the clock input signal when the clock inhibit and reset terminals are both low. Nine of the ten decoded outputs are low, with the remaining one high, at any given time. The outputs go high sequentially, in phase with the clock signal, with the selected output remaining high for one full clock cycle. An additional carry out signal completes one cycle for every ten clock input cycles, and can be used to ripple-clock additional 4017s in multidecade counting circuits.

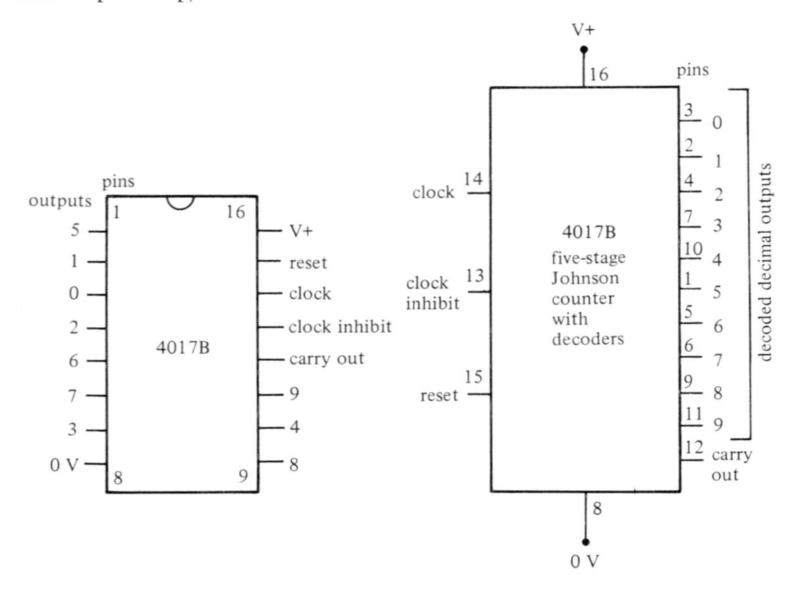

Figure 8.1 *Outline and functional diagram of the 4017B decade counter*

The 4017B counting cycle can be inhibited by setting the clock inhibit terminal (pin 13) high. A high signal on the reset terminal (pin 15) clears the counters to zero and sets the decoded 0 output terminal high. In use, all unused inputs of the IC must be tied high or low, as appropriate. The IC has a built-in Schmitt trigger on its clock line, and is thus not fussy about clock-signal rise and fall times.

4017B applications

Figures 8.3 to *8.6* show some basic ways of using the 4017B in counting applications. In *Figure 8.3* the IC is connected as a decade counter/divider in which the output frequency is one-tenth of the input clock frequency. Here, the reset and inhibit terminals are grounded and the output is taken from the carry out terminal, the decoded outputs being ignored.

Figure 8.4 shows how three decade dividers of the *Figure 8.3* type can be cascaded to make a three-decade divider that generates outputs at 1/10th, 1/100th and 1/1000th of the clock frequency; the carry out signal of each counter is used to provide the clock signal to the following stage. Note that in this particular circuit the four outputs are buffered via simple CMOS inverters (made from 4001 gates etc.), to ensure that output loading does not

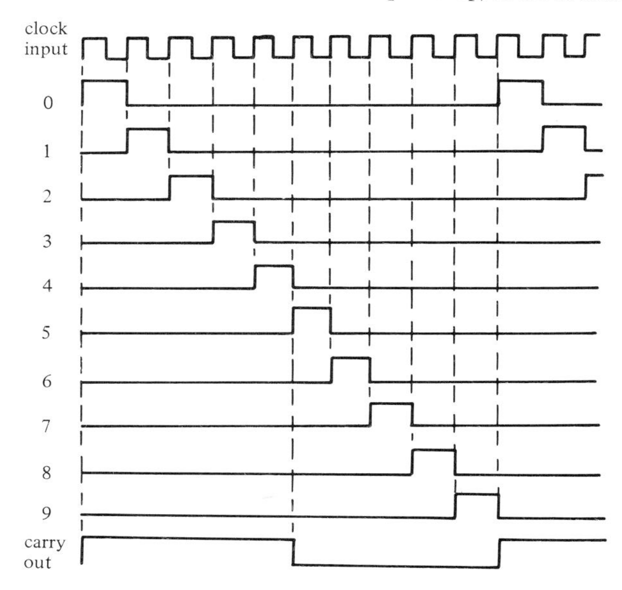

Figure 8.2 *Waveform timing diagram of the 4017B, with its reset and clock inhibit terminals grounded*

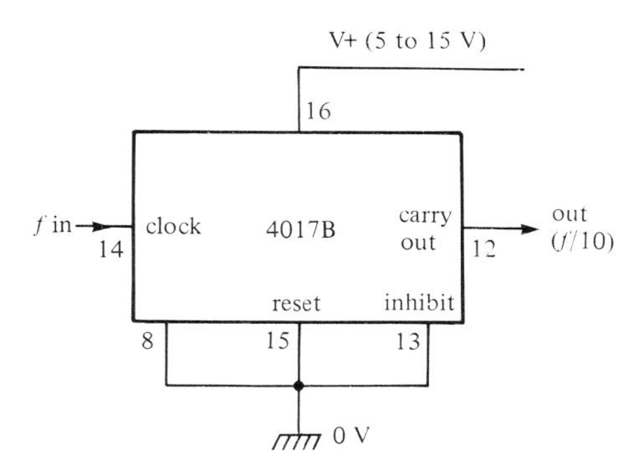

Figure 8.3 *4017B connected as a decade counter/divider*

degrade the clock signal rise times. Thus, if the clock input signal is derived from a 1 MHz crystal oscillator, the circuit can be used as a laboratory frequency standard, generating frequencies of 1 MHz, 100 kHz, 10 kHz, and 1 kHz.

Figure 8.5 shows a simple method of connecting the 4017B as a divide-by-N counter (N = any whole number from two to nine) with N decoded outputs.

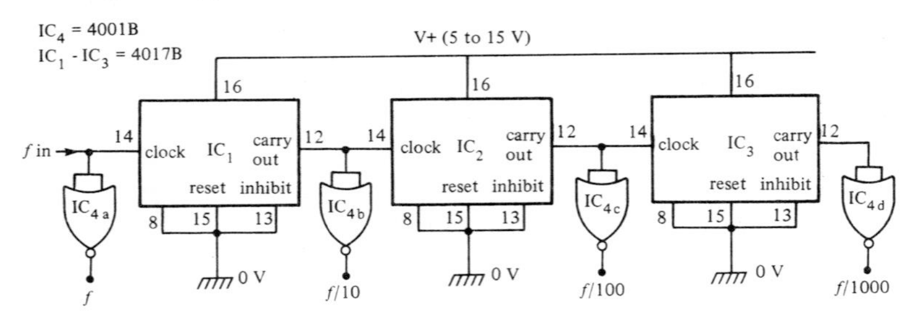

Figure 8.4 *Any number of 4017Bs can be cascaded to make a multidecade divider. If the outputs are externally loaded they should be buffered, as shown here*

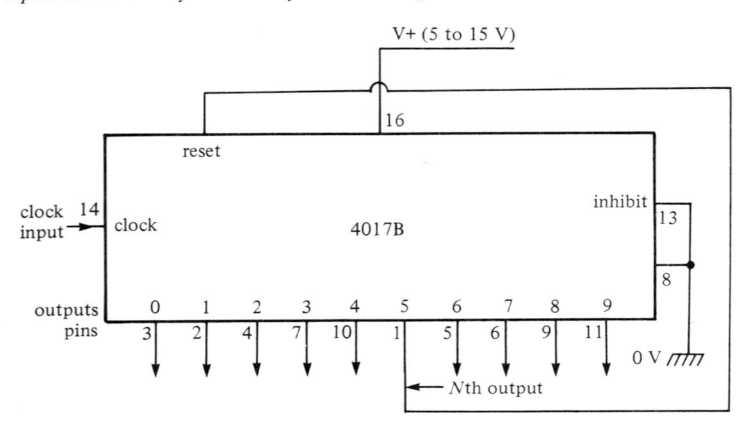

Figure 8.5 *Simple method of connecting the 4017B as a divide-by-*N *(*N *= two to nine) counter with* N *decoded outputs. The circuit is shown set for divide-by-five operation*

Here, the *N*th decoded output is simply short circuited to the reset terminal so that the counter resets to zero on the arrival of the *N*th clock pulse. In the diagram, the circuit is set for divide-by-five operation.

Figure 8.6 shows a slightly more sophisticated version of the divide-by-*N* counter, in which logic gates control the reset operation via the IC_{1a}-IC_{1b} flip-flop. The operation here is such that the reset command is given on the arrival of the *N*th clock pulse and is maintained while the clock pulse remains high, but is removed automatically when the clock pulse goes low again.

Sequencing

The most important feature of the 4017B is the fact that it provides up to ten fully decoded outputs, making the IC ideal for use in a whole range of

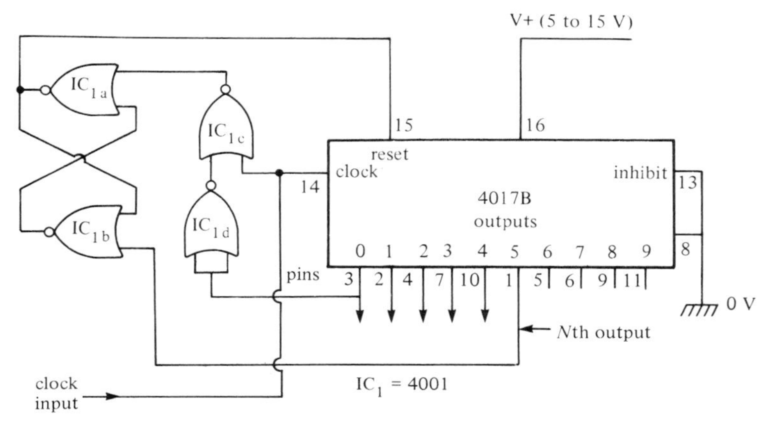

Figure 8.6 *Alternative method of connecting the 4017B as a divide-by-*N *counter. The circuit is shown set for divide-by-five operation*

sequencer applications in which the outputs are used to drive LED displays, relays, sound generators, etc. *Figures 8.7* and *8.8* show ways of extending the usefulness of the IC in such applications.

Figure 8.7 shows how to connect the IC so that it stops operating after completing a predetermined counting sequence. In the diagram the counter is set to stop when its clock inhibit terminal is driven high by the 9 output, but the counter can in fact be inhibited via any one of the 4017's decoded output terminals. The count sequence can be restarted by pressing reset button PB_1.

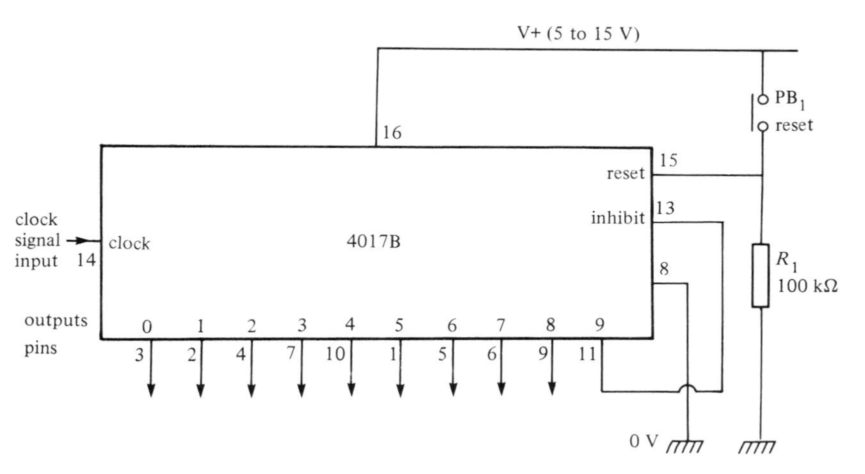

Figure 8.7 *Method of connecting the 4017B for sequence-and-stop action*

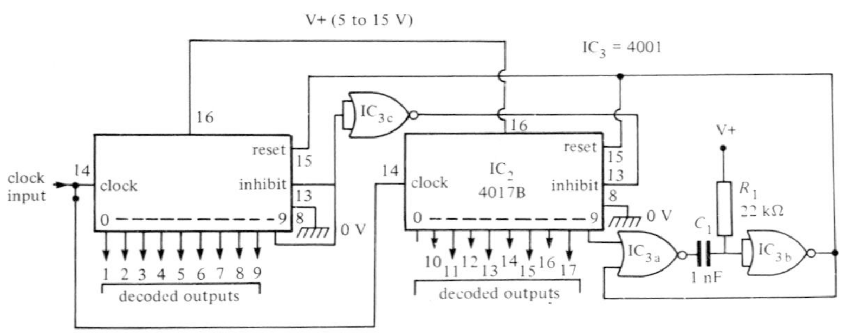

Figure 8.8 *A ten- to seventeen- stage counter/decoder, set for divide-by-seventeen operation*

Finally, *Figure 8.8* shows how a pair of 4017Bs can be connected to give up to seventeen stages of fully decoded counting. The clock input signal is fed to both ICs, but when the count is below nine the 9 output of IC_1 is low and causes the clock inhibit terminal of IC_2 to be set high via IC_{3c}, so IC_2 is not influenced by the clock signals. When the ninth clock pulse arrives the 9 output of IC_1 goes high and inhibits IC_1 from further clocking action; simultaneously, the clock inhibit terminal of IC_2 is driven low via IC_{3c}, enabling IC_2 to respond to subsequent clock signals. Eventually, on the arrival of the seventeenth clock pulse, the 9 output of IC_2 goes momentarily high and triggers the IC_{3a}-IC_{3b} 15 μs monostable, which resets both counters to their 0 states. The counting sequence then repeats.

Note that the 9 output of IC_1 and the 0 and 9 outputs of IC_2 are lost in the counting action, so the circuit gives a maximum of seventeen usable counter/divider stages. The circuit can be made to count by any number in the range ten to seventeen by connecting the free terminal of IC_{3a} to the appropriate output terminal of IC_2.

The 4022B octal counter

The 4022B can be regarded as an octal (divide-by-eight) brother of the 4017B. It is a four-stage walking-ring synchronous counter with eight fully decoded outputs, so arranged that they go high sequentially. *Figure 8.9* shows the outline and functional diagram of the device. For normal octal counting, the reset and inhibit terminals are tied low; the carry out signal completes one cycle for every eight input clock cycles. The IC has a built-in Schmitt trigger on its clock input line, and is thus not fussy about clock-signal rise and fall times. The clock is inhibited by a high signal on pin 13; the counter is reset by a high signal on pin 15; and the counter advances on each positive transition of the clock signal.

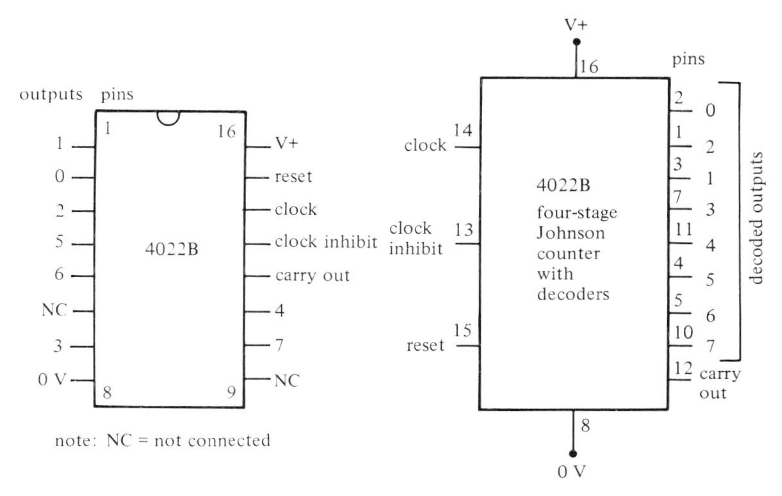

Figure 8.9 *Outline and functional diagram of the 4022B octal counter*

Synchronous up counters

Synchronous up counters are normally used in fairly simple applications in which it is necessary to merely count a number of input pulses, or divide them by a fixed ratio, and (perhaps) then display the results on a seven-segment LED or LCD readout unit.

Three basic families of CMOS up counters are readily available, to suit virtually all possible needs. The oldest family comprises the 4026B and 4033B types, which are decade counters with built-in decoders which give seven-segment display outputs that can directly drive sensitive LED displays. Another family of devices comprises the 4518B and 4520B duals, which house two counters in each sixteen-pin package. The 4518B is a dual decade counter with BCD outputs, and the 4520B is a dual hexadecimal (divide-by-sixteen) counter with a four-bit binary output. The outputs of these ICs must be decoded externally if they are to be used to drive seven-segment displays etc.

The third family of up counters comprises the 40160B to 40163B range of presettable counters, which can be made to reset to (or start to count from) either zero or any four-bit number that is fed into a set of four preset (jam) pins. The 40160B and 40162B are decade dividers, and the 40161B and 40163B are binary dividers.

Let's now look at details of each of these three families of up counters, starting with the 4026B and 4033B types.

4026B and 4033B counters

The 4026B and the 4033B are synchronous up counters which incorporate decoding circuitry that gives a seven-segment output suitable for directly driving a sensitive seven-segment common-cathode LED display; the output drive currents are limited to only a few milliamperes. Both IC's have clock, clock inhibit, and reset input terminals, and a carry output terminal that completes one cycle for every ten input clock cycles and can be used to clock following decades in a counting chain. The counters are advanced on the positive transition of the input clock cycle.

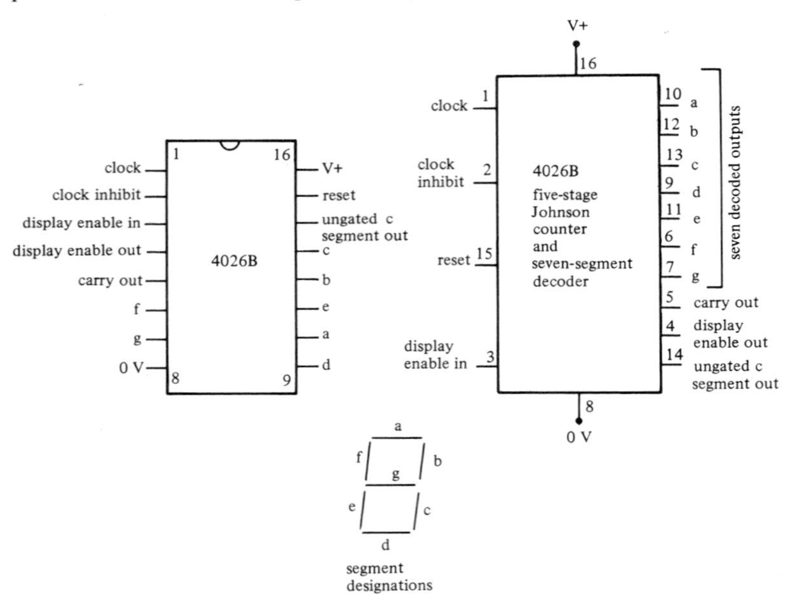

Figure 8.10 *Outline and functional diagram of the 4026B decade counter with seven-segment display driver and display enable control*

Figure 8.10 shows the outline and functional diagram of the 4026B. This IC features display enable input and output terminals, which enable the entire display to be blanked. Normally, the display enable in pin is held high; the display blanks (but the IC continues to count) when this pin is pulled low. The IC also has an ungated *c* segment output terminal, which can be used with external logic to make the IC count by numbers other than ten.

Figure 8.11 shows the outline and functional diagram of the 4033B. This IC features ripple blanking input and output (RBI, RBO) terminals, which can be used to automatically blank leading and trailing zeros in multidecade

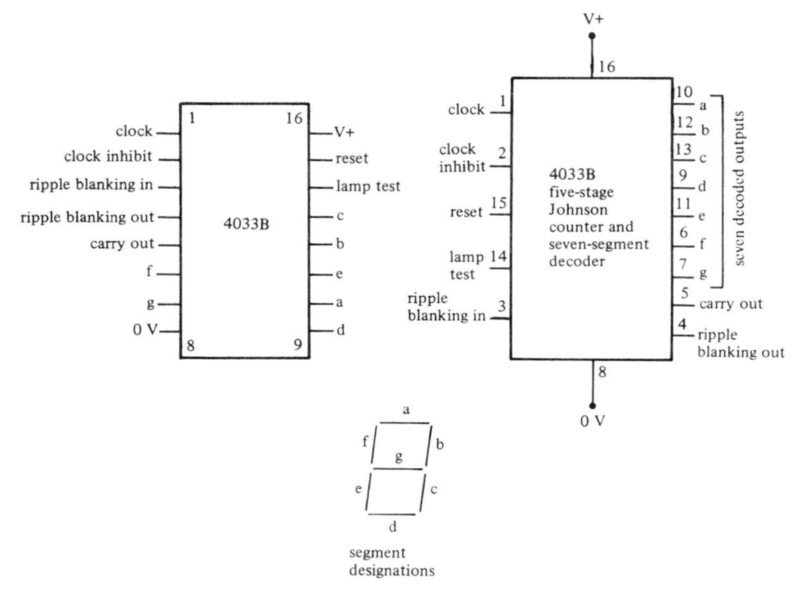

Figure 8.11 *Outline and functional diagram of the 4033B decade counter with seven-segment display driver and ripple blanking facility*

applications; the 0 display blanks automatically when the ripple blanking input pin is held low. The IC also features a lamp test pin, which is normally held low but which drives all seven decoded outputs high when the input pin is taken high.

Figure 8.12 shows how to connect the 4026B and the 4033B for simple decade counter/display operation. In both cases the reset and clock inhibit terminals are tied low. In the case of the 4026B, the display enable input must be tied high if the display is to be illuminated. In the case of the 4033B, the

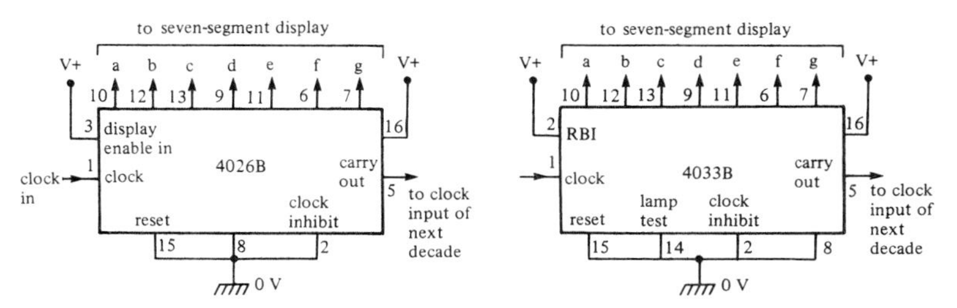

Figure 8.12 *Method of connecting the 4026B and 4033B for normal decade dividing/display operation. In the case of the 4033B, the RBI connection does not give suppression of display zeros*

ripple blanking input terminal must be tied high if the display is required to give normal operation, or must be tied low if it is required to give zero suppression. Note that in both circuits, if multidecade counting is to be used, the carry out of one stage must be used to provide the clock input signal of the next stage.

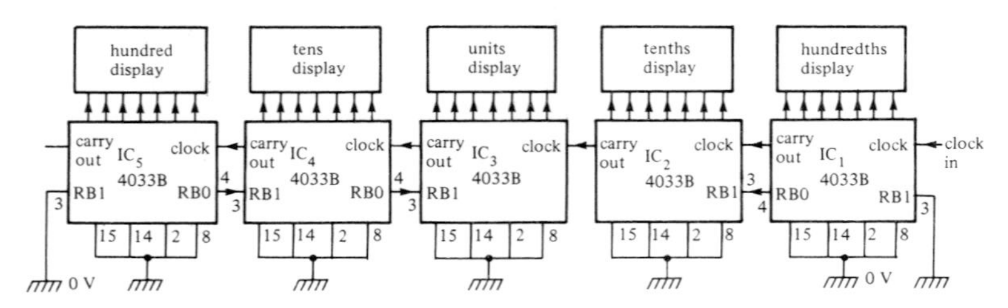

Figure 8.13 *Method of connecting the RBI and RBO terminals of the 4033B to give leading and trailing zero suppression in a multidecade display so that, for example, the count 009.90 is displayed as 9.9*

Finally, *Figure 8.13* shows how to interconnect several 4033Bs to give automatic suppression of leading and trailing zeros so that, for example, the count 009.90 will actually be displayed as 9.9. To get leading zero suppression (on the integer side) the RBI (pin 3) terminal of the most significant digit (MSD) counter must be tied low, and its RBO (pin 4) terminal must be taken to the RBI terminal of the next MSD counter, and so on down to the units counter. To get trailing zero suppression (on the fraction side of the display) the RBI of the least significant bit (LSB) must be tied low, and its RBO terminal must be taken to the RBI terminal of the next LSB counter, and so on, to the first counter in the fractions chain.

When contemplating use of the 4026B or the 4033B, note that these ICs do *not* incorporate data latches; consequently, the displays tend to blur when the ICs are actually going through a counting cycle.

4518B and 4520B counters

The 4518B and 4520B are dual up counters with binary-coded outputs. The 4518B is a dual decade counter with binary-coded decimal (BCD) outputs, and the 4520B is a dual hexadecimal (divide-by-sixteen) counter with four-bit binary outputs. The ICs have identical outlines and functional diagrams, as shown in *Figure 8.14*.

An unusual feature of these counters is that they can be clocked on either the positive or the negative edge of the clock signal. For positive-edge clocking, feed the clock to the clock terminal and tie the enable terminal high.

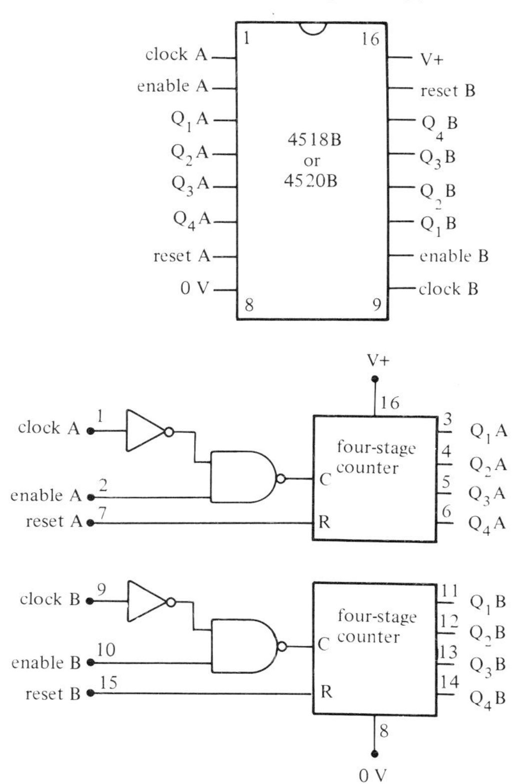

Figure 8.14 *Outline and functional diagrams of the 4518B (decade) and 4520B (binary) dual counters*

For negative-edge clocking, feed the clock to the enable terminal and tie the clock terminal low. The counters are cleared by a high level on their reset pins, and clear asynchronously.

Note that these counters do not have a carry output; to cascade counter stages, the negative-edge clocking feature must be utilized, as shown in *Figure 8.15*. Here the Q_4 output of each counter is fed to the enable input of the following stage, which must have its clock terminal tied low.

40160B to 40163B counters

This range of counters comprise presettable types, in which the four-bit output can be made to agree with a four-bit code set up on four preset pins,

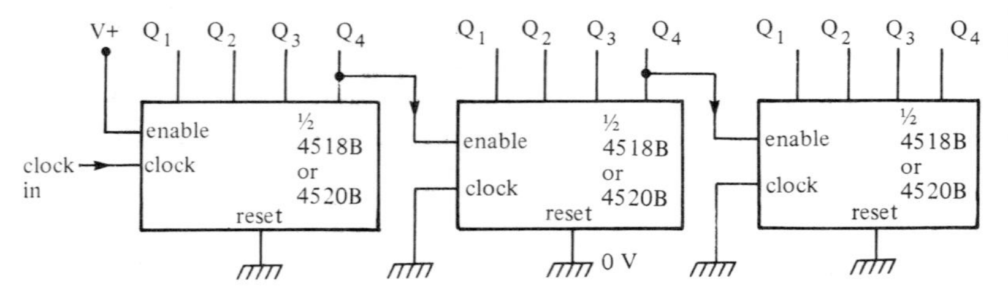

Figure 8.15 *Methods of cascading 4518B or 4520B counters for ripple operation*

regardless of other conditions. This facility is useful, for example, in producing a count/display system that will repeatedly count from 1 to x, rather than from 0 to x as in a non-programmable counter.

The 40160B to 40163B ICs have identical outlines and functional diagrams, as shown in *Figure 8.16*, but differ in their methods of counting and clearing. The 40160B and 40162B are both decade counters with BCD outputs, but the 40160B has an asynchronous clear action whereas the 40162B has a synchronous clear action. The 40161B and 40163B are both hexadecimal counters with four-bit outputs, but the clear action of the 40161B is asynchronous whereas that of the 40163B is synchronous.

Thus, a low level on the $\overline{\text{clear}}$ terminal of the 40160B or 40161B asynchronous types immediately sets all four outputs low, regardless of the states of the clock, $\overline{\text{load}}$, or enable inputs, but in the case of the 40162B or

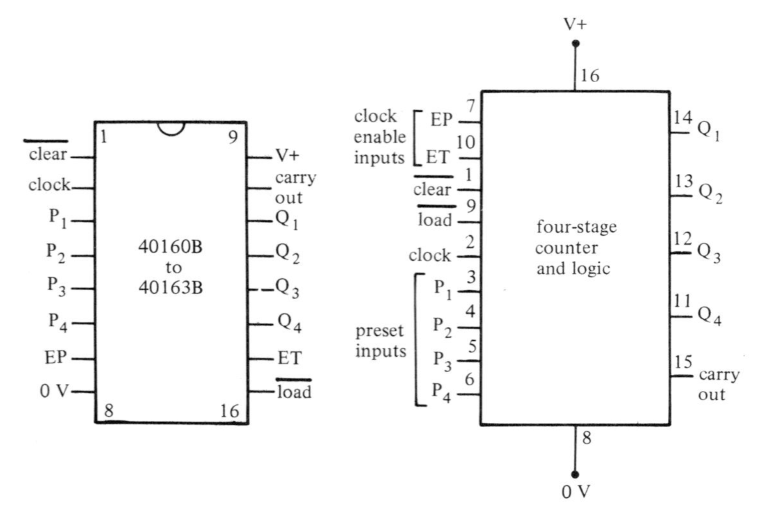

Figure 8.16 *Outline and functional diagrams of the 40160B to 40163B range of programmable four-bit counters*

40163B synchronous types a low level on the $\overline{\text{clear}}$ terminal sets the outputs low only on the arrival of the *next* positive clockedge. Note that a low level on the $\overline{\text{load}}$ terminal disables the counters and causes the outputs to agree with the preset (P) data on the arrival of the *next* clock pulse, regardless of the stages of the enable inputs.

The counters have two clock enable pins (EP and ET), which must be tied high for normal counting operation. These pins are available to facilitate an internal carry look-ahead feature that is useful in fast counting applications. *Figure 8.17* shows how to connect the ICs for use as normal counters.

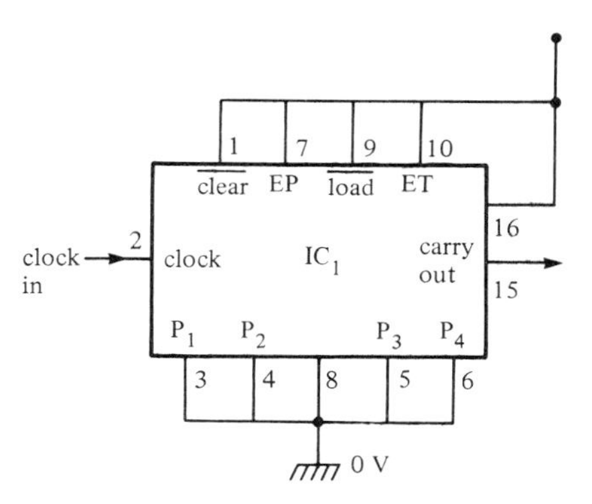

Figure 8.17 *Method of connecting the 40160B to 40163B range of ICs for normal counter operation*

Synchronous up/down counters

Synchronous up/down counters are devices that can be made to count in either direction, either by using a single clock input in conjunction with a direction control, or by using two seperate clock signals, with one controlling the up counting and the other controlling the down count. Up/down counters are very versatile devices; they can be used in the fixed mode as conventional up counter/display units or as count-down counter/display devices. In the active mode, they can be used as difference or add/subtract counters or bidirectional counter/display units etc.

Three basic types or families of CMOS up/down counters are readily available. They are all versatile presettable types, giving undecoded four-bit binary or BCD outputs which can be forced to agree with a four-bit binary number set on four jam or preset terminals.

The oldest type of CMOS up/down counter is the 4029B. This device uses a single clock signal, with count direction controlled via an up/down terminal. The IC can act as either a decade (BCD output) counter, or a binary (four-bit binary output) counter, depending on the setting of a binary/decade select terminal.

The 4510B and 4516B family of up/down counters also uses a single clock signal, with count direction controlled via an up/down terminal, but in this case the count scale is fixed. The 4510B is a decade counter with BCD outputs, and the 4516B is a binary counter with a four-bit binary output.

Finally, the 40192B and 40193B family of up/down counters use seperate clocks to give the up and down counting action. The 40192B is a decade counter with BCD outputs, and the 40193B is a binary counter.

Let's now look in detail at each of these three sets of up/down counters, starting with the 4029B type.

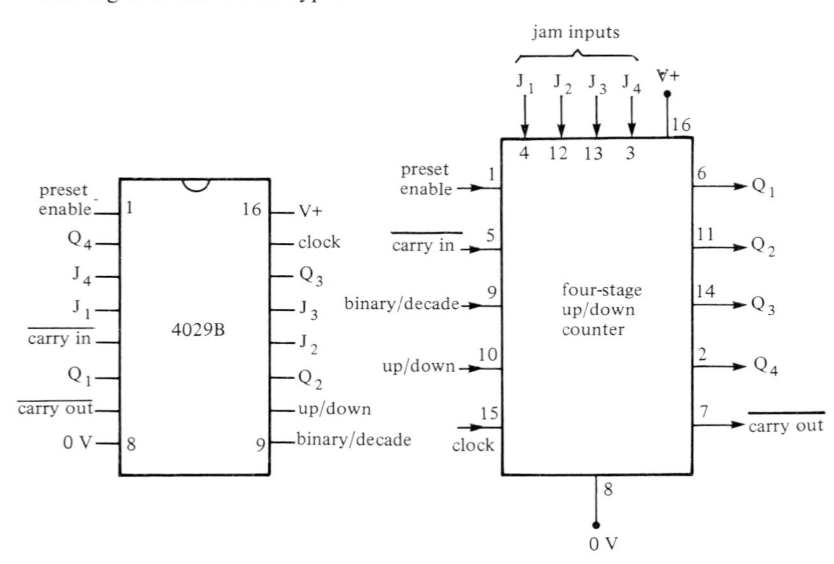

Figure 8.18 *Outline and functional diagram of the 4029B presettable up/down counter*

The 4029B up/down counter

Figure 8.18 shows the outline and functional diagram of this versatile presettable IC; a high preset enable signal forces the outputs to immediately (asynchronously) agree with the binary code set on the jam inputs. The IC acts as a binary counter when the binary/decade input terminal is high, or as a decade counter (with BCD outputs) when the terminal is low. The IC counts up when the up/down input terminal is high, or down when the terminal is low. Note, however, that the up/down control must *only be changed when the clock signal is positive*.

The 4029B terminal marked carry in is actually a clock disabling terminal, and is held low for normal clocking operation. When carry in and preset enable are low, the counter shifts (up or down) one count on each positive transition of the clock signal. The carry out terminal of the IC is normally high and goes low only when the counter reaches its maximum count in the up mode or its minimum count in the down mode (provided that the carry in signal is low).

The actions of the carry in and carry out terminals of the 4029B are designed to facilitate fully synchronous action in multidecade counting applications, as shown in *Figure 8.19*, where all ICs are clocked in parallel and the carry out terminal of each counter is used to enable the following one (at a 'one-tenth of' rate) via its carry in terminal.

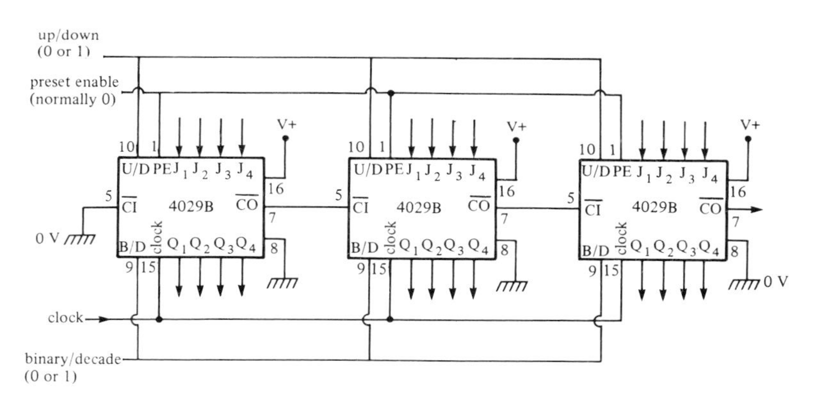

Figure 8.19 *Method of cascading 4029Bs for synchronous parallel clocking*

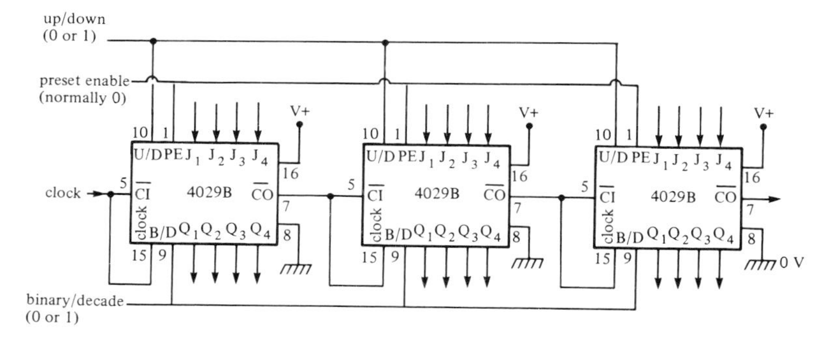

Figure 8.20 *Method of cascading 4029Bs for asynchronous ripple clocking*

Numbers of 4029Bs can also be cascaded and clocked in the asynchronous ripple mode by using the connections shown in *Figure 8.20*, but in this case the Q outputs of different counters will not be glitch-free when decoded. Note that in this diagram the clock and $\overline{\text{carry in}}$ terminals are joined together on each IC, to ensure that false counting will not occur if the up/down input is changed during a terminal count.

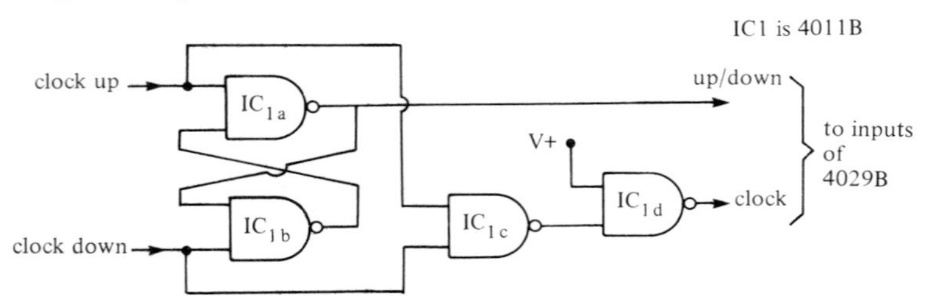

Figure 8.21 *Simple circuit for converting the 4029B to a dual-clock up/down counter*

Finally, *Figure 8.21* shows a simple circuit that can be used to convert the 4029B into a dual-clock up/down counter, in which clock signals on the up terminal give up clocking, and clock signals on the down terminal give down clocking. The counter shifts (up or down) on the positive transition of either clock signal. Note that only one clock terminal can be used at a time, and that the unused terminal must be held high.

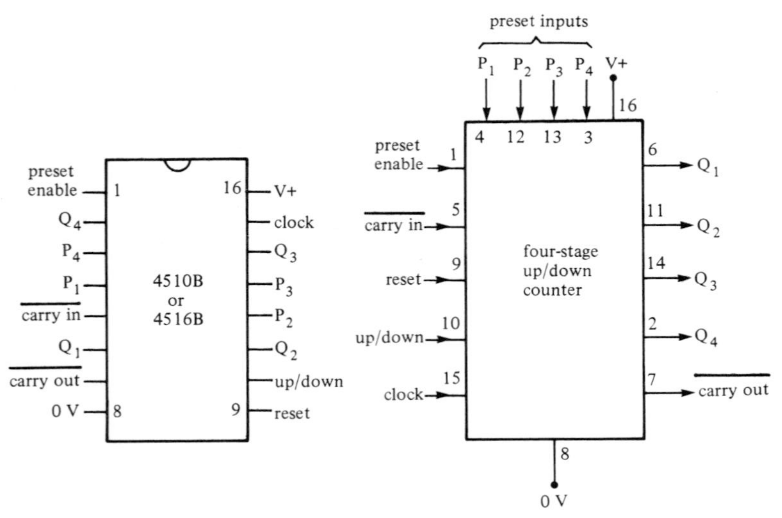

Figure 8.22 *Outline and functional diagram of the 4510B (decade) and 4516B (binary) up/down counters*

4510B and 4516B up/down counters

The 4510B and 4516B are presettable up/down counters. The 4510B is a decade counter with BCD outputs, and the 4516B is a binary (hexadecimal) counter. The ICs have identical outlines and functional diagrams, as shown in *Figure 8.22*. These counters have a reset terminal which enables all outputs to be set low (cleared) by applying a high level on the reset line. The outputs can be made to agree with a binary code on the preset (P) terminals by applying a high level on the preset enable line. The IC counts up when the up/down terminal is high, and down when the terminal is low; the up/down input must only change when the clock signal is high.

The $\overline{\text{carry in}}$ terminal is actually a clock disabling terminal, and is held low for normal operation. When $\overline{\text{carry in}}$ (and reset and preset enable) is low, the counter shifts one count on each positive transition of the clock. The $\overline{\text{carry out}}$ terminal is normally high and goes low only when the counter reaches its maximum count in the up mode or its minimum count in the down mode (provided that the $\overline{\text{carry in}}$ terminal is low).

Numbers of 4510Bs and 4516Bs can be cascaded and clocked in parallel to give fully synchronous action by using the connection shown in *Figure 8.23*,

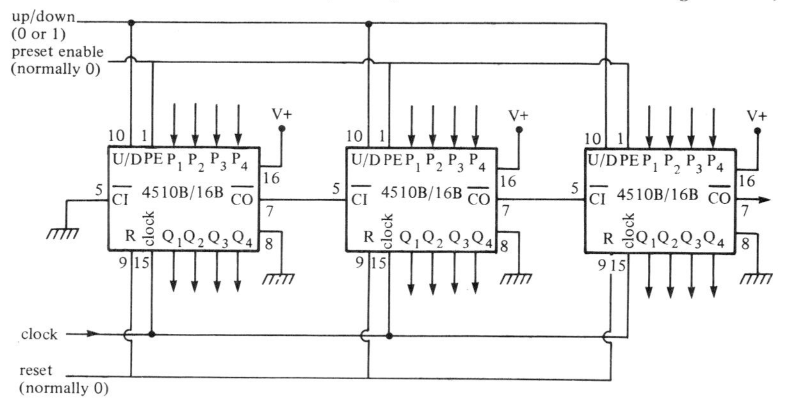

Figure 8.23 *Method of cascading 4510B or 4516B ICs for synchronous parallel clocking*

or can be cascaded in the asynchronous ripple mode by using the connections shown in *Figure 8.24*, which ensures counting validity even if the up/down input is changed during a terminal count.

40192B and 40193B up/down counters

The 40192B and 40193B are presettable dual-clock up/down counters. The 40192B is a decade counter with BCD outputs, and the 40193B is a

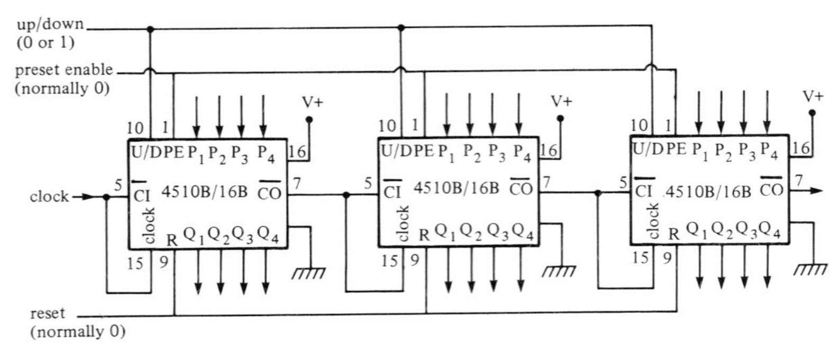

Figure 8.24 *Method of cascading 4510B or 4516B ICs for asynchronous ripple clocking*

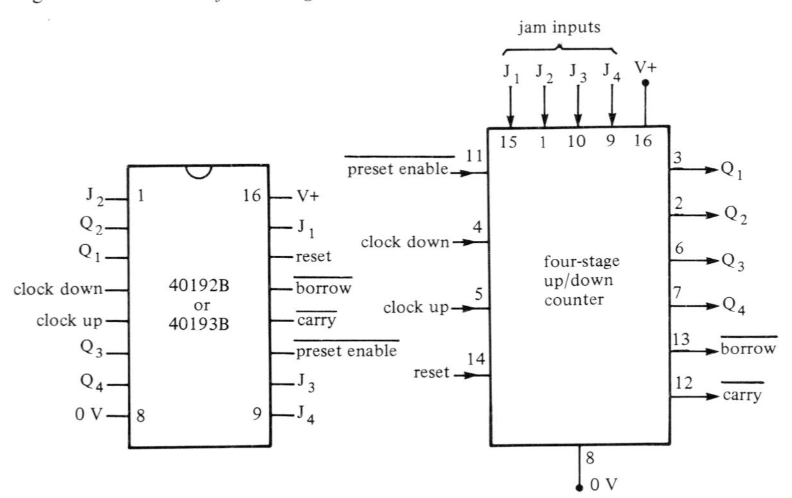

Figure 8.25 *Outline and functional diagram of the 40192B (decade) and 40193B (binary) dual-clock up/down counters*

hexadecimal counter. The ICs have identical outlines and functional diagrams, as shown in *Figure 8.25*. A high on the reset terminal asynchronously sets all four outputs to zero. A low on the $\overline{\text{preset enable}}$ terminal asynchronously sets the outputs to the values set on the four jam (J) input terminals.

The counters have two clock input lines, one controlling the up count and the other controlling the down count. Only one clock input terminal must be used at a time, and the unused input must be tied high. The counter shifts one count (up or down) on each positive transition of the used clock line.

The $\overline{\text{carry}}$ and $\overline{\text{borrow}}$ output signals of the counters are normally high, but the $\overline{\text{carry}}$ signal goes low one-half clock cycle after the counter reaches its maximum count in the up mode, and the $\overline{\text{borrow}}$ output goes low one-half clock cycle after the counter reaches minimum count in the down mode.

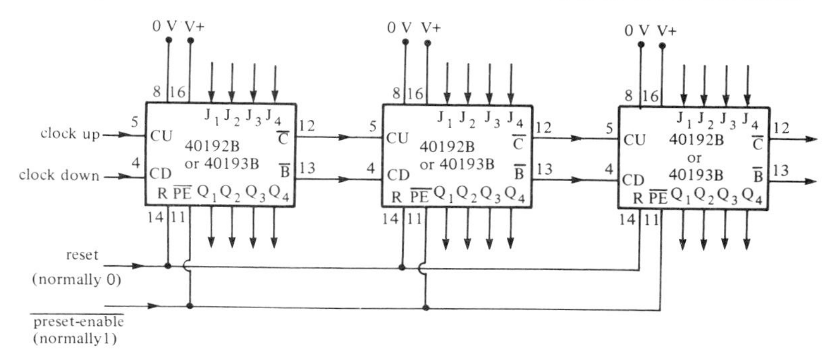

Figure 8.26 *Method of cascading 40192B or 40193B counters*

Finally, *Figure 8.26* shows how to connect 40192Bs or 40193Bs in cascaded multiple IC applications. The $\overline{\text{borrow}}$ and $\overline{\text{carry}}$ outputs of each IC are simply connected directly to the clock down and clock up inputs respectively of the following ICs.

9 Down counters and decoders

In the previous two chapters we have looked in detail at CMOS D and JK flip-flops, and at various sophisticated types of CMOS up and up/down counter/divider ICs. In this chapter we round off our look at CMOS counter/dividers and associated devices with a detailed examination of dedicated presettable down counters, plus a brief survey of various types of decoders and display-driver ICs.

Presettable down counter basics

Dedicated presettable down counters are special ICs that can be externally programmed to divide by any number by simply feeding that number (in binary or BCD form) to a set of preset terminals. *Figure 9.1* illustrates the basic principle. Here, a programmable decade down counter (PDDC) has the BCD code 6 fed to its four preset inputs via a BCD thumbwheel switch, so that the counter automatically gives divide-by-six operation.

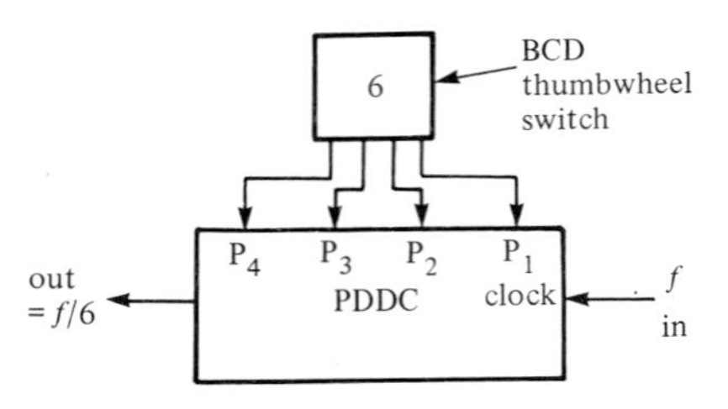

Figure 9.1 *A programmable decade down counter (PDDC) can count/divide by any BCD number fed to its preset terminals*

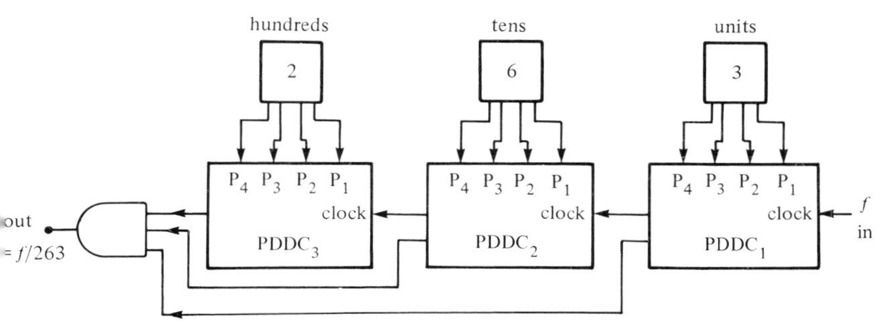

Figure 9.2 *PDDCs have programmable cascadability; the output of this circuit is equal to the sum of the individual division ratios*

The *unique* feature of the PDDC, however, is that it has a characteristic known as 'programmable cascadability', as illustrated in *Figure 9.2*. Here, the hundreds counter is set to divide-by-two, the tens counter to divide-by-six, and the units counter to divide-by-three, and the overall system gives a division ratio of $200 + 60 + 3 = 263$. Conventional counters (*Figure 9.3*) would, of course, give an overall division ratio of $200 \times 60 \times 3 = 36000$ if cascaded with the same division ratios.

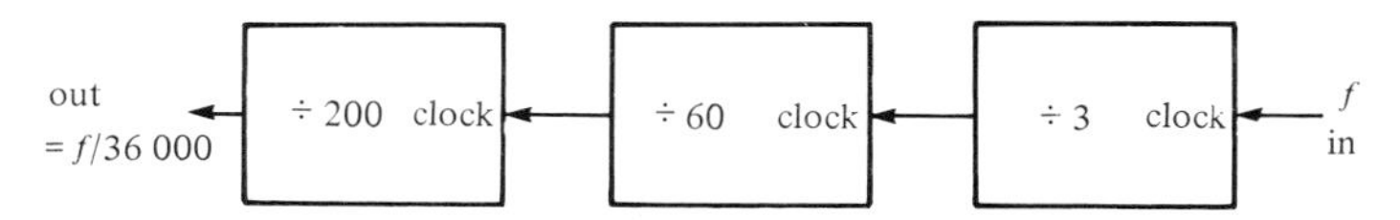

Figure 9.3 *When conventional counters are cascaded they give a final output equal to the product of the individual division ratios*

Thus, the very special feature of the presettable down counter is its programmable cascadability, which enables it to be made into systems which count or divide by any desired numbers that are fed into its preset inputs. Practical PDDCs can readily be made to act as programmable frequency dividers or frequency synthesizers, or as programmable down counters or timers. The programming can be done electromechanically via thumbwheel switches, or electronically via microprocessor control etc.

How down counters work

Figure 9.4 shows the basic functional diagram of a decade version of the presettable down counter (binary versions are similar). The unit houses a four-stage synchronous down counter which shifts down one count on each positive transition of the clock signal, as indicated by the truth table of *Figure*

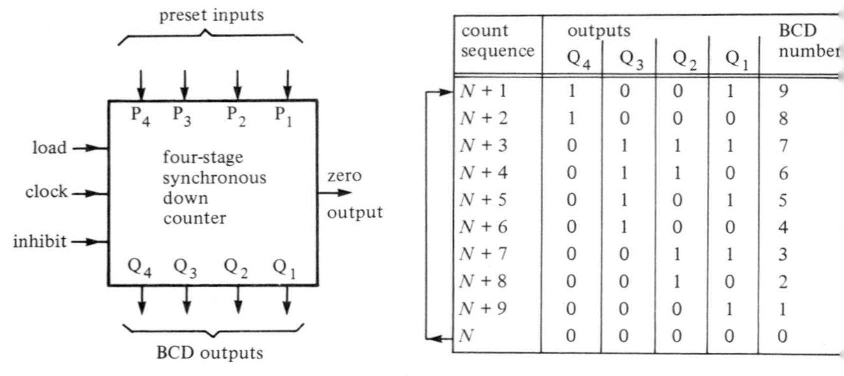

count sequence	outputs Q_4	Q_3	Q_2	Q_1	BCD number
$N+1$	1	0	0	1	9
$N+2$	1	0	0	0	8
$N+3$	0	1	1	1	7
$N+4$	0	1	1	0	6
$N+5$	0	1	0	1	5
$N+6$	0	1	0	0	4
$N+7$	0	0	1	1	3
$N+8$	0	0	1	0	2
$N+9$	0	0	0	1	1
N	0	0	0	0	0

Figure 9.4 *Basic functional diagram of a presettable decade down counter, together with the truth table of the actual counter stages*

9.4. Note in particular that if a count sequence starts with all stages in the 0 (minimum count) state, the first arriving clock pulse causes the counters to jump to the BCD 9 (code 1001) or maximum count state. The outputs of the four counters are available at the Q terminals.

The clock signal of the PDDC can be disabled by a high level on the inhibit control. A decoded zero output is available and goes high only when all counters are in the 0 or minimum count state. Four preset input terminals are provided, and the counters can be forced to take up the BCD states set on these by applying a high level to the load pin.

Figure 9.5 shows how to connect a PDDC as a programmable counter that counts down from the BCD number that is loaded via the start button, and stops counting and gives a high output when that number of clock pulses has been received. Suppose that the BCD number 6 is loaded via the start button,

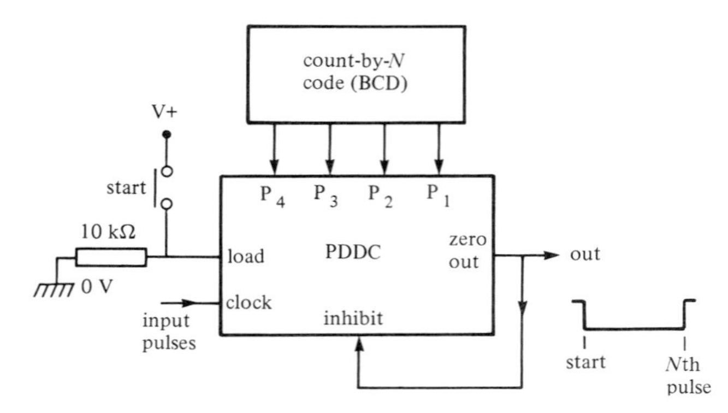

Figure 9.5 *PDDC connected as a programmable down counter*

and clock signals are then initiated. On the arrival of each clock pulse the IC counts down one step, going through the numbers 5, 4, 3, 2, 1 and finally, on the arrival of the sixth pulse, to 0, at which point the zero output terminal goes high and activates the inhibit terminal, causing any further clock pulses to be ignored. The count sequence is then complete. It can be restarted by pressing the start button.

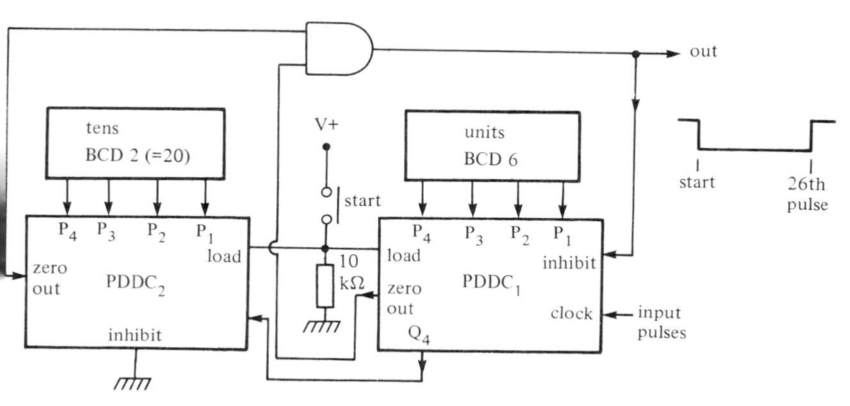

Figure 9.6 *Two-decade programmable down counter, set for count 26 operation*

Figure 9.6 shows how two PDDCs can be cascaded to make a two-decade programmable down counter giving a count of 26. Note that the zero outputs of the ICs are ANDed to generate the inhibit control of $PDDC_1$ (and also the final output), and that the clock signal of $PDDC_2$ comes from the Q_4 output of $PDDC_1$. The count action of the circuit is as follows.

When the start button is pressed the BCD number 2 is loaded into the tens counter and 6 is loaded into the units counter; the clock input signal is then initiated. $PDDC_1$ counts down from 6 to 0 through the first six clock pulses but then, since the zero outputs are not both high at this stage, simply starts acting as a divide-by-ten counter and starts counting down from the BCD number 9 (code 1001) on the arrival of the seventh pulse, simultaneously sending a single clock pulse to $PDDC_2$ as Q_4 switches high. Ten pulses later (on the seventeenth pulse) $PDDC_1$ sends another clock pulse to $PDDC_2$, making its zero output go high, and nine pulses later (on the 26th pulse) the zero output of $PDDC_1$ also goes high, at which point the output of the AND gate goes high and inhibits further counting action. The sequence is then complete.

Figure 9.7 shows how to wire a single PDDC as a programmable timer with an output that goes high as the start button is pressed but then goes low again a preset time later. The circuit is like that of *Figure 9.5*, except that its final output is inverted and the clock signal is taken from a fixed time-reference

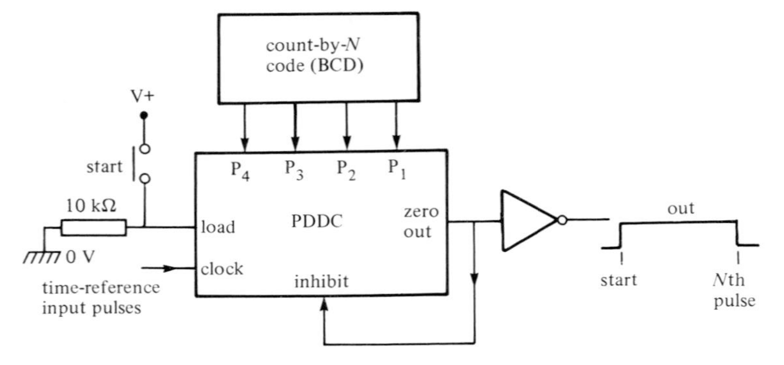

Figure 9.7 *PDDC connected as a programmable timer*

source (e.g. 1 pulse/second). Note that the *Figure 9.6* circuit can also be made to act as a programmable timer by similarly inverting its final output and taking the clock signal from a fixed time-reference source.

Frequency division

Figure 9.8 shows how to wire a PDDC as a programmable frequency divider. The divide-by-N code is fed to the preset terminals, and the output is taken from the zero terminal, which is coupled back to the load pin. Suppose that at the start of the count the BCD number 4 has been loaded into the counter. On the arrival of the first clock pulse the counter decrements to 3, on the second pulse to 2, on the third to 1, and on the fourth to 0, at which point the zero output goes high and loads the BCD number 4 back into the counters, so the

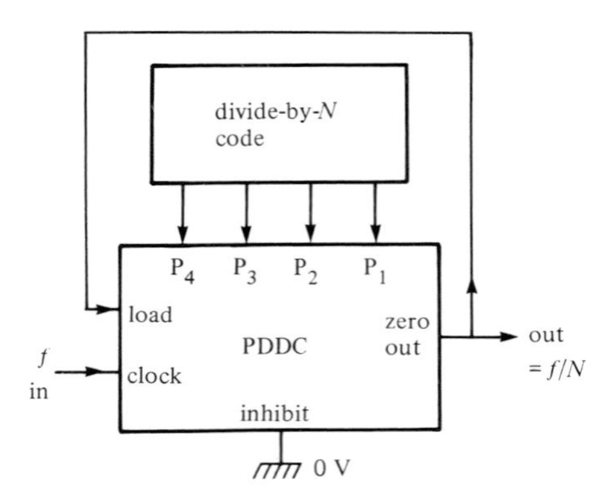

Figure 9.8 *PDDC connected as a programmable frequency divider*

whole sequence starts over again and the zero output goes back low. Thus, the counter repeatedly counts by the number (4) set on the preset inputs, and the output (from the zero terminal) takes the form of a narrow pulse with a width of a few hundred nanoseconds.

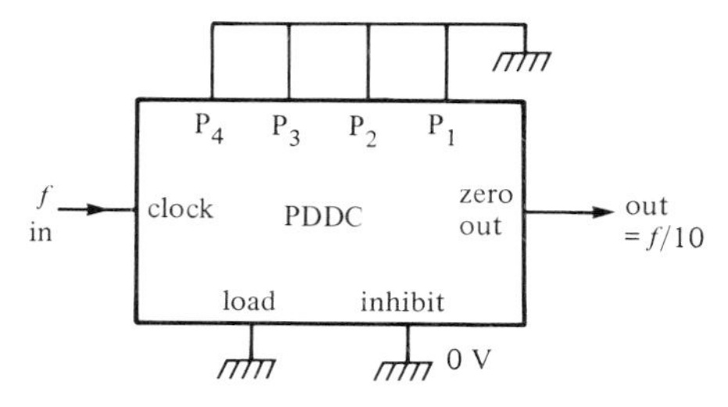

Figure 9.9 *PDDC connected as a simple decade frequency divider*

Figure 9.9 shows a PDDC connected as a simple divide-by-ten counter/divider. Here, the load terminal is grounded, so the preset codes have no effect and the counter repeatedly cycles through its basic BCD count, from 9 to 0 and then back to 9 again, and so on. The output, taken from the zero output terminal, goes high for one full clock cycle in every ten.

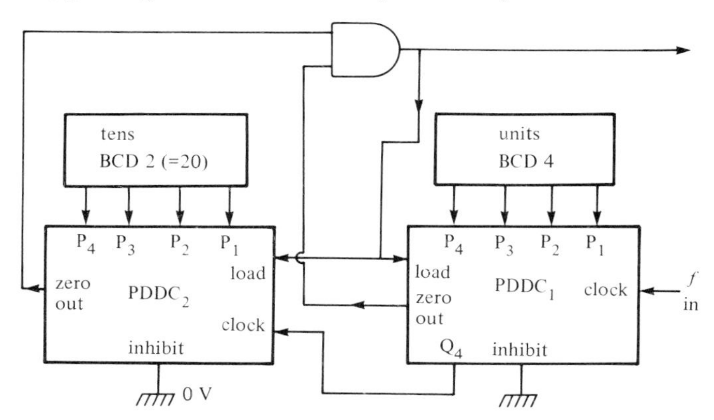

Figure 9.10 *Two-decade programmable frequency divider, set for divide-by-24 operation*

Figure 9.10 shows how to wire two PDDCs to give cascaded programmable frequency divider operation. In this case the units counter is set for divide-by-four and the tens for divide-by-two operation (= ÷24). The two zero outputs are ANDed to give the load action (and also the final output), and the $PDDC_2$ clock signal is derived from the Q_4 output of $PDDC_1$. The circuit operates as follows.

Assume that at the start of the count cycle the BCD number 24 is loaded into the counters. For the first few counts in the cycle $PDDC_1$ counts from 4 to 0 and then, since $PDDC_2$ is not also in the 0 state at this moment, goes into the normal divide-by-ten mode, jumping to the 9 state (and simultaneously feeding a clock pulse to $PDDC_2$ as Q_4 switches high) and counting down from there. This action continues until eventually, on the arrival of the 24th pulse, the zero outputs of both ICs go high together, at which point the output of the AND gate goes high and loads the BCD number 24 back into the counters, and the whole sequence starts over again.

Thus, the *Figure 9.10* circuit repeatedly divides by the two-decade number that is programmed in, producing a narrow (a few hundred nanoseconds) outputs pulse from the AND gate on the completion of each 'divide-by-24' counting cycle.

Frequency synthesis

The main application of programmable frequency dividers is in programmable frequency synthesis, where they are used in conjunction with a phase-locked loop (PLL) as shown in *Figure 9.11*. Here, the output of a wide-range voltage-controlled oscillator (VCO) is fed, via the programmable divide-by-N counter, to one input of a phase detector, which has its other input taken from a fixed-frequency crystal-controlled reference generator. The phase detector produces an output voltage proportional to the difference between the two input frequencies; this voltage is filtered and fed back to the VCO control input in such a way that the VCO automatically self-adjusts to

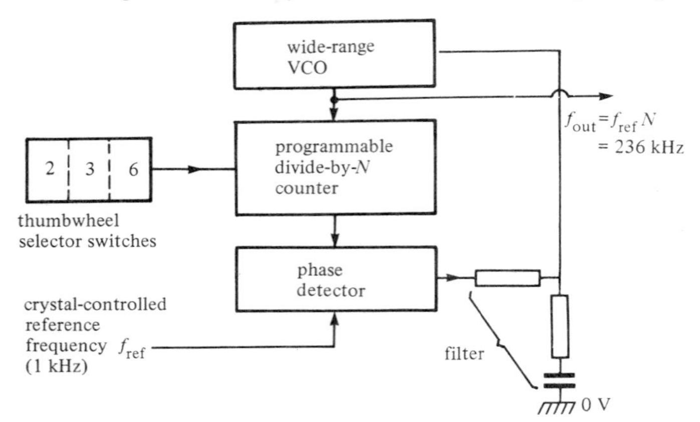

Figure 9.11 *A programmable divide-by-*N *counter can be used in conjunction with a PLL to make a precision programmable frequency synthesizer*

bring the variable input frequency of the phase detector to the same value as the reference frequency, at which point the PLL is said to be 'locked'.

Note that the output frequency of the VCO is N times the value of the frequency on the variable input of the phase detector. Thus, when the PLL is locked, the VCO output frequency is N times the reference frequency, e.g. if $N = 236$ and $f_{ref} = 1$ kHz, f_{out} equals 236 kHz and has crystal accuracy.

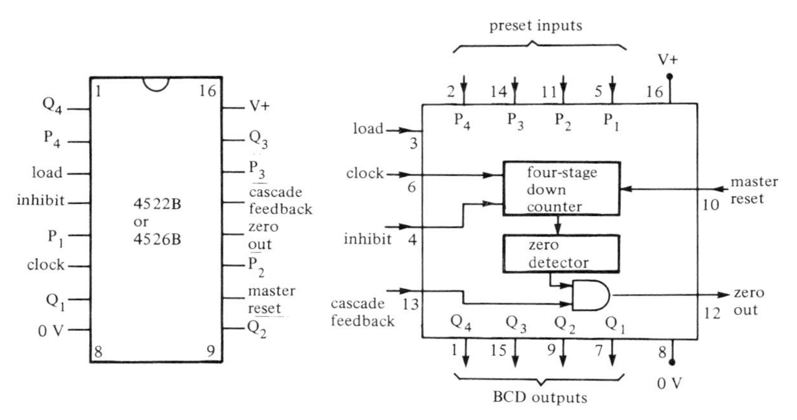

Figure 9.12 *Outline and functional diagram of the 4522B (decade) and 4526B (binary) programmable down counters*

The 4522B and 4526B

The best known family of CMOS programmable cascadable down counters comprises the 4522B (decimal) and the 4526B (binary) four-bit ICs, which both have the outlines and functional diagrams shown in *Figure 9.12*. Functionally, these ICs are almost identical to the basic circuit of *Figure 9.4*, except that the counters can be synchronously reset to the zero state by taking the master reset pin high, and that an AND gate is built into the zero output line, so that the zero output can only go high if the cascade feedback terminal is also high, thus enabling cascading to be achieved without the use of external gates.

The 4522B and 4526B can be used in the same ways as shown in *Figures 9.5* to *9.11*, except that the master reset pin is normally grounded and the AND gate is built into the ICs. When the ICs are used alone, the cascade feedback (CF) terminal must be tied high to enable the zero output. When cascading two or more ICs, tie the zero output of the most significant digit (MSD) package to the CF terminal of the MSD – 1 package, repeating the process on all less significant dividers except the first. *Figure 9.13* shows practical

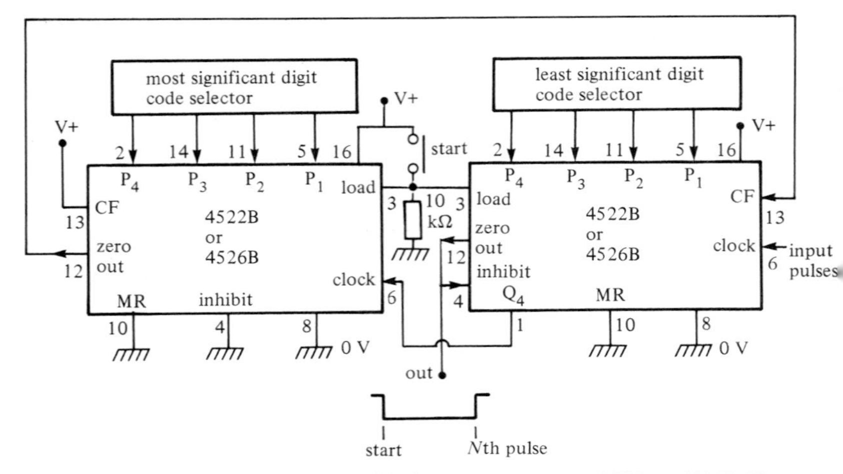

Figure 9.13 *Two-stage programmable down counter, using 4522B or 4526B ICs*

connections for making a two-stage programmable down counter, and *Figure 9.14* shows the connections for a two-stage programmable frequency divider.

When using these ICs, note that all unused inputs (including presets) must be tied high or low, as appropriate, and that the outputs of all internal counter stages are available via the Q terminals, enabling the counter states to be decoded via external circuitry.

The 40102B and 40103B

The 40102B and 40103B are another popular family of CMOS programmable cascadable down counter ICs, but in this case each device effectively houses a *pair* of presettable four-bit down counters cascaded in a single package, with only the zero output (which goes low under the zero-count condition) of the counters externally available; Q outputs are not provided. *Figure 9.15* shows the outline, functional diagram, and truth table that is common to both ICs. The 40102B functions as a two-decade BCD-type down counter, and the 40103B as an eight-bit (or two four-bit words) binary type. Both types clock down on the positive transition of the clock signal.

Codes that are applied to the eight preset pins of the ICs can be loaded asynchronously by pulling the $\overline{\text{AL}}$ pin low, or synchronously (on the arrival of the next clock pulse) by pulling the $\overline{\text{SL}}$ pin low. When the $\overline{\text{clear}}$ input is pulled low the counter asynchronously clears to its maximum count. When the inhibit control is pulled high it inhibits both the clock counting action and the zero output action, thereby acting as a carry in terminal in cascaded applications.

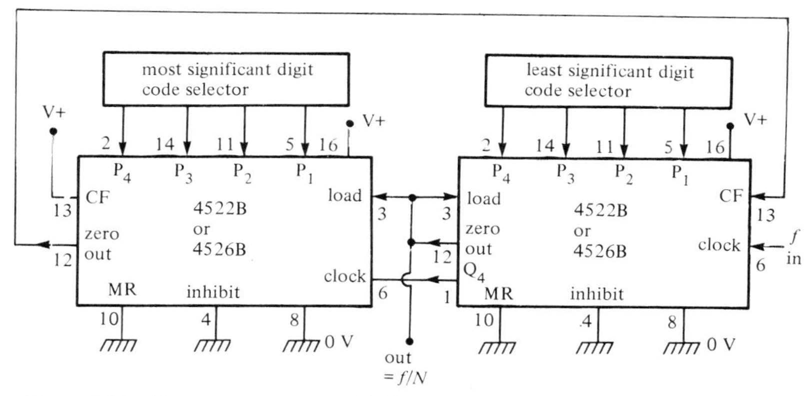

Figure 9.14 *Two-stage programmable frequency divider, using 4522B or 4526B ICs*

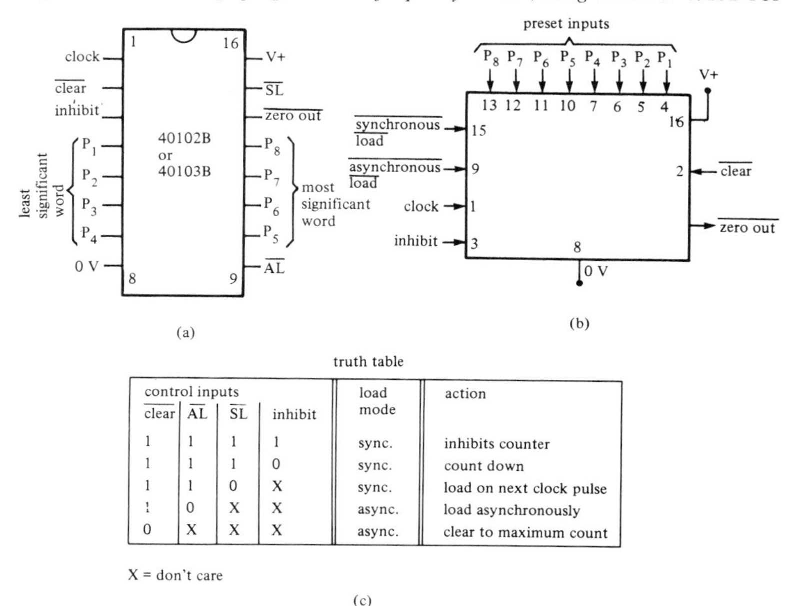

control inputs				load mode	action
$\overline{clear}$	$\overline{AL}$	$\overline{SL}$	inhibit		
1	1	1	1	sync.	inhibits counter
1	1	1	0	sync.	count down
1	1	0	X	sync.	load on next clock pulse
1	0	X	X	async.	load asynchronously
0	X	X	X	async.	clear to maximum count

X = don't care

(c)

Figure 9.15 *(a) Outline, (b) functional diagram, and (c) truth table of the 40102B two-decade and 40103B eight-bit binary down counters*

Figures 9.16 to *9.19* show four basic ways of using this family of presettable down counters. *Figure 9.16* shows the connections for making a programmable eight-bit or two-word timer or down counter, and *Figure 9.17* shows the circuit of a programmable frequency divider. The latter circuit gives

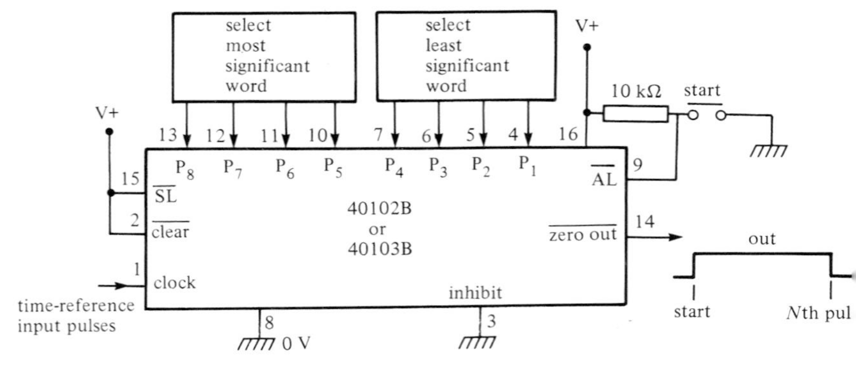

Figure 9.16 *Programmable timer using a 40102B or 40103B*

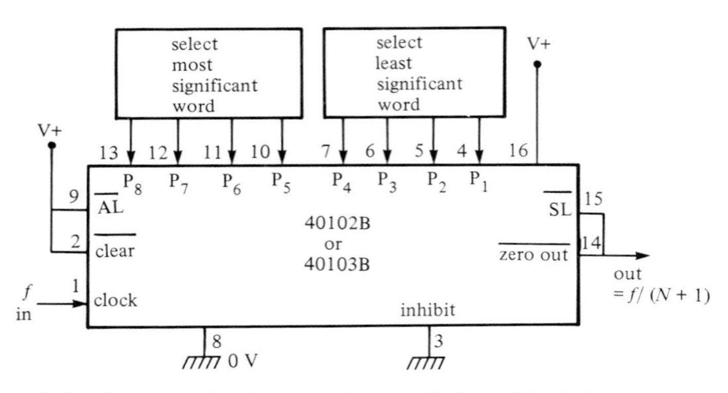

Figure 9.17 *Programmable frequency divider (divide-by-(N+1))*

divide-by-($N+1$) operation, its output going low for one full clock cycle under the zero-count condition; true divide-by-N operation can be obtained by tying $\overline{\text{SL}}$ high and wiring $\overline{\text{zero out}}$ to $\overline{\text{AL}}$, but in this case the output pulses will have widths of only a few hundred nanoseconds.

Finally, *Figures 9.18* and *9.19* show the basic connections that are used to cascade 40102B or 40103B stages in large-word programmable applications. The *Figure 9.18* connections give ripple operation, and the *Figure 9.19* connections give fully synchronous operation (for fast applications).

Decoders

Most of the counter/dividers that we have looked at in this and the previous two chapters give four-bit coded outputs, which take the standard form

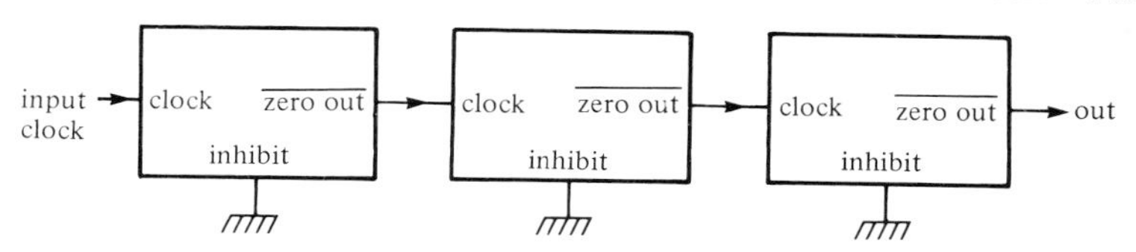

Figure 9.18 *Method of ripple cascading 40102B or 40103B counters*

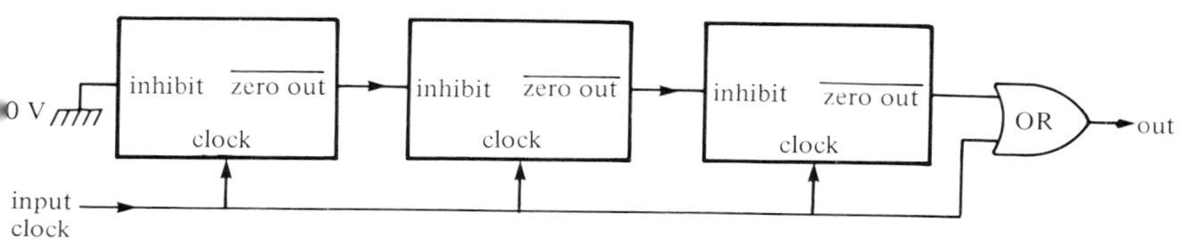

Figure 9.19 *Method of synchronous cascading 40102B or 40103B counters*

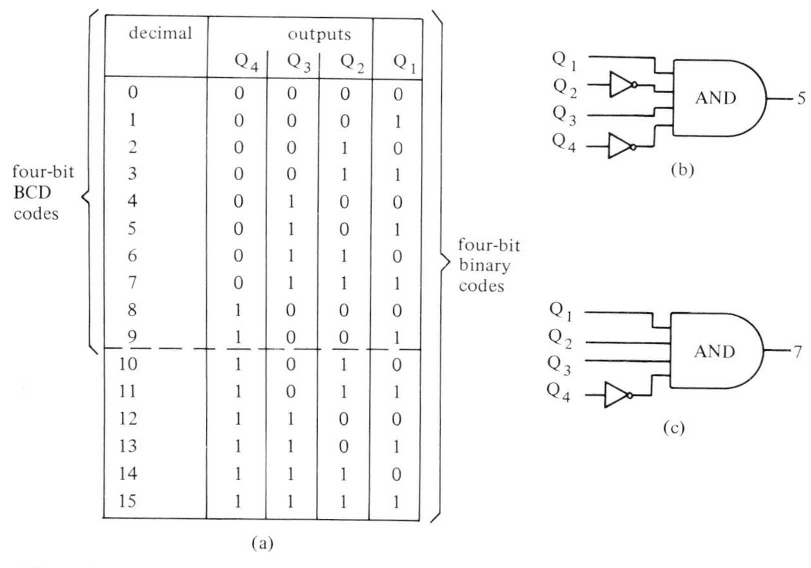

decimal	outputs Q_4	Q_3	Q_2	Q_1
0	0	0	0	0
1	0	0	0	1
2	0	0	1	0
3	0	0	1	1
4	0	1	0	0
5	0	1	0	1
6	0	1	1	0
7	0	1	1	1
8	1	0	0	0
9	1	0	0	1
10	1	0	1	0
11	1	0	1	1
12	1	1	0	0
13	1	1	0	1
14	1	1	1	0
15	1	1	1	1

Figure 9.20 *The coded output states of a four-bit counter are shown in (a), and (b) and (c) show how to decode the numbers 5 and 7, respectively*

shown in *Figure 9.20(a)*. Thus, when the counters are in the 5 state they give an output code of 0101, and in the 7 state give 0111 (read with Q_4 to the left).

Individual output states of the counters can easily be decoded and used for driving external display units, control lines, etc. by using the basic logic techniques shown in *Figure 9.20 (b)* and *9.20(c)*. Here outputs that are high (logic-1) in the desired code state are fed directly to the inputs of a four-input AND gate, and those that are low are fed to the input of the gate via inverter stages. If BCD numbers are to be decoded, and the numbers lie between 2 and 7 inclusive, the Q_4 code can be ignored and three-input AND gating can be used.

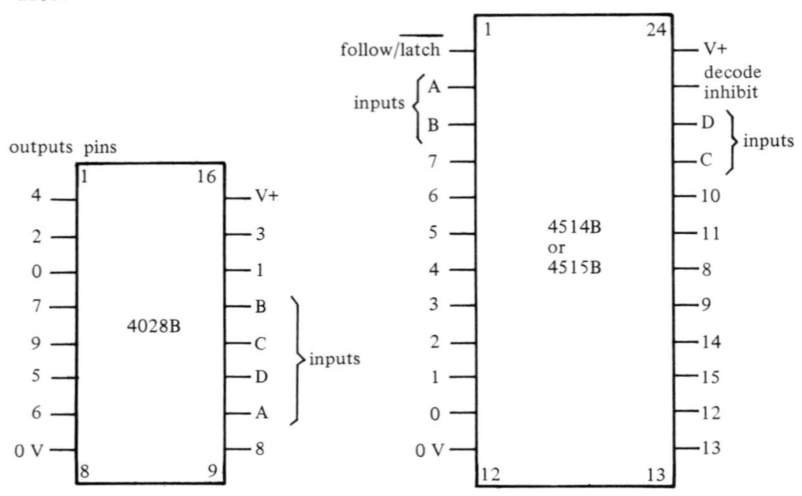

Figure 9.21 *Outlines and pin notations of the 4028B BCD-to-decimal (one-of-ten) decoder, and of the 4514B (active-high output) and 4515B (active-low output) four-bit binary (one-of-sixteen) decoders*

If more than two of three code states are to be decoded, it is usually economic to use dedicated CMOS decoder ICs, such as the 4028B, 4514B, or 4515B (*Figure 9.21*) for the purpose. The 4028B BCD-to-decimal decoder simply gives direct decoding of the ten BCD or binary numbers 0 to 9 inclusive; input terminal A corresponds to Q_1 of the code, and C corresponds to Q_4 of the code.

The 4514B and 4515B are full four-bit decoders, with an individual output for each of the sixteen possible code numbers. The 4514B gives active-high outputs (all outputs except the selected one are normally low), and the 4515B gives active-low outputs. These ICs are far more sophisticated than the 4028B type, and have follow/$\overline{\text{latch}}$ or F/$\overline{\text{L}}$ control on pin 1 and a decode inhibit or DI control on pin 15.

The DI control disables all decoding functions; it drives all outputs low in the 4514B, or high in the 4515B, irrespective of the states of all other pins. The ICs act as straight or direct decoders when the F/$\overline{\text{L}}$ terminal is high, but when the terminal is pulled low it latches the prevailing input code into memory and retains it, irrespective of the subsequent states of the input code, as long as F/$\overline{\text{L}}$

remains high; the latched code is decoded and fed to the output in the normal way.

BCD-to-seven-segment decoders/drivers

The BCD outputs of decade counters can easily be decoded and used to drive seven-segment LED or LCD displays by using suitable decoder/driver ICs. Seven-segment displays have the standard format and terminal notations shown in *Figure 9.22(a)*, with the segment pins individually available on the package.

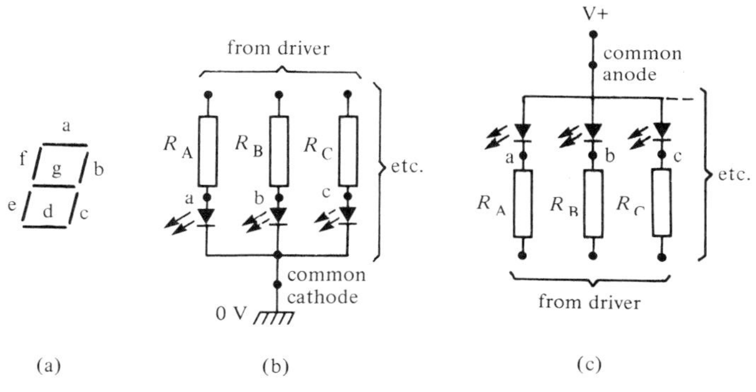

Figure 9.22 *Seven-segment LED or LCD displays have the standard format shown in (a). LED types are available in (b) common-cathode or (c) common-anode forms. Current-limiting resistors must be wired in series with the segment pins*

LED seven-segment displays are available in either common-cathode or common-anode forms, and *Figures 9.22(b)* and *9.22(c)* show small sections of the equivalent circuits of these types. Common-cathode types must be driven by ICs that can source significant currents, and common-anode types must be activated by devices that can sink significant currents. Note that in both types a current-limiting resistor must be wired in series with each segment pin.

The most popular CMOS IC for driving seven-segment LED displays is the 4511B BCD-to-seven-segment latch/decoder/LED-driver, which has the pin notations and functional diagram shown in *Figure 9.23*. This IC is ideally suited to driving common-cathode displays, since its outputs can each source up to 25 mA. The 4511B is an easy IC to use. It has four BCD input terminals, seven segment-driving output terminals, and only three input control terminals.

The $\overline{\text{lamp test}}$ input terminal is normally tied high; when pulled low it turns on all seven segments of the display, irrespective of the input code. The $\overline{\text{blank}}$

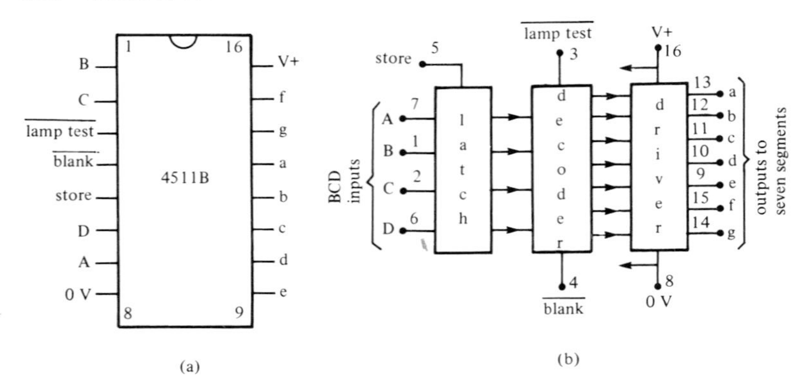

Figure 9.23 *(a) Outline and (b) functional diagram of the 4511B BCD-to-seven-segment latch/decoder/LED-driver*

input terminal is also normally tied high; when pulled low it turns off all display segments, irrespective of the input code. The store control enables the IC to give either direct or latched decoder operation. When store is low, the IC gives direct decoding of the BCD inputs. If store is switched high, the BCD input that is present at the moment of switching is latched into the internal store and then decoded, and this BCD number is held (and the actual BCD input is ignored) as long as store remains high.

Seven-segment LCD drivers

Seven-segment liquid crystal displays (LCDs) have the same format as LED types (*Figure 9.22(a)*), except that their common terminal is known as the backplane or BP. LCDs must effectively be driven by AC signals with virtually zero DC components. In practice, the AC signal takes the form of a square wave with a frequency in the range 30 Hz to 200 Hz.

Older LCD drivers relied on the use of dual power supplies to give the AC drive. Modern types use the single-supply bridge technique of *Figure 9.24* to give the necessary AC drive. Here, when a segment is turned on, the segment and BP are driven by anti-phase square waves, as shown. Note here that, as far as the display is concerned, it is the value of the segment voltage *relative to the backplane voltage* that is important. Thus, in part A of the waveform the segment is 10 V positive to BP, and in part B it is 10 V negative to BP, so the LCD is effectively driven by an AC signal with a peak-to-peak value of 20 volts and with zero DC value. In this system, when a segment is turned off it is simply shortcircuited to BP.

The most popular CMOS IC for driving seven-segment LCDs is the 4543B

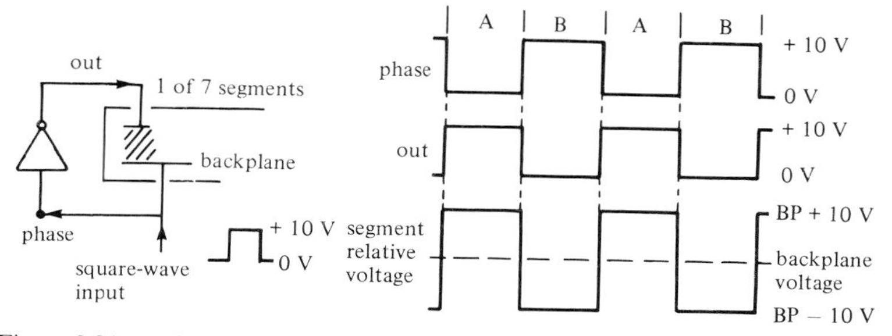

Figure 9.24 *Voltage-doubling bridge method of driving liquid crystal displays*

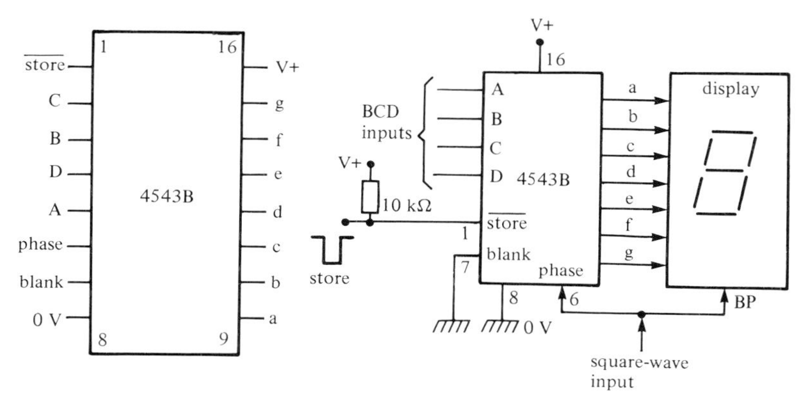

Figure 9.25 *Outline and usage diagram of the 4543B BCD-to-seven-segment latch/decoder/LCD-driver*

BCD-to-seven-segment latch/decoder/LCD-driver, which uses the techniques described above. *Figure 9.25* shows the outline and basic usage diagram of the device, which must have its phase terminal connected to the backplane of the display and driven from a symmetrical external square wave. The $\overline{\text{store}}$ control gives direct decoding when pulled high, or gives latched decoding when pulled low. The blank terminal is normally grounded, and blanks the entire display when pulled high.

Conclusion

In this and the previous two chapters we have looked at a selection of CMOS counting and decoding ICs. *Tables 9.1* to *9.7* list the actual devices that we have looked at. In practice, all of these devices are intended for use in fairly simple one-off or limited-run applications. For dedicated applications such as making frequency meters or digital clocks, special VLSI (very large-scale integration) ICs are readily available. Similarly, when driving seven-segment LCD or LED displays of more than three sections, it is often economical to use a dedicated VLSI IC with full multiplexing facilities to do the job; these ICs are also readily available.

Table 9.1 *Non-synchronous (ripple) counters*

Type	Description
4020B	Fourteen-stage (divide-by-16384) counter
4024B	Seven-stage (divide-by-128) counter
4040B	Twelve-stage (divide-by-4096) counter
4060B	Fourteen-stage counter, with built-in oscillator

Table 9.2 *General-purpose counters*

Type	Description
4013B	Dual type D flip-flop
4017B	Decade counter with ten decoded outputs
4018B	Decade divide-by-*N* counter
4022B	Octal counter with eight decoded outputs
4027B	Dual JK flip-flop

Table 9.3 *Synchronous up counters*

Type	Description
4026B	Decade counter with seven-segment display outputs (features display enable control)
4033B	Decade counter with seven-segment display outputs (feature ripple blanking)
4518B	Dual decade counter, BCD outputs
4520B	Dual hexadecimal counter, four-bit binary outputs
40160B	Presettable decade counter with asynchronous clear
40161B	Presettable binary counter with asynchronous clear
40162B	Presettable decade counter with synchronous clear
40163B	Presettable binary counter with synchronous clear

Table 9.4 *Synchronous presettable up down counters*

Type	Description
4029B	Four-bit binary or BCD decade counter
4510B	BCD decade counter
4516B	Four-bit binary counter
40192B	BCD decade dual-clock counter
40193B	Four-bit binary dual-clock counter

Table 9.5 *Synchronous presettable down counters*

Type	Description
4522B	Decade counter
4526B	Binary counter
40102B	Two-decade BCD counter
40103B	Eight-bit binary counter

Table 9.6 *Four-bit decoders*

Type	Description
4028B	BCD-to-decimal decoder
4514B	One-of-sixteen decoder, active-high outputs
4515B	One-of-sixteen decoder, active-low outputs

Table 9.7 *BCD-to-seven-segment decoder drivers*

Type	Description
4511B	Latch/decoder/LED-driver
4543B	Latch/decoder/LCD-driver

10 Circuit miscellany

To conclude the volume, this final chapter presents a miscellaneous collection of useful CMOS digital IC circuits that do not readily fit into any of the specific classifications of earlier chapters. These circuits include alarm-call generators, security alarms, DC lamp controllers, and some unusual designs based on the 4046B phase-locked-loop IC.

Alarm-call generators

Inexpensive CMOS gate ICs such as the 4001B can easily be used (in conjunction with one or more transistors) to make a variety of audible-output alarm-call generator circuits. These generators may give monotone, pulsed-tone, or warble-tone outputs, may give latching, non-latching, or 'one-shot' operation, and can be designed to give output powers ranging from a few milliwatts up to about 18 watts. In this section we look only at low-power versions of these alarm-call generations; power-boosting circuits are described in a later section.

Figure 10.1 shows the practical circuit of a low-power switch-activated 800 Hz (monotone) alarm call generator. Here, two gates of a 4001B are wired as a gated astable multivibrator (see Chapter 5) with a frequency set at 800 Hz by the R_1-C_1 values, and the remaining two gates of IC are disabled by taking their inputs to ground. The action of this astable is such that it is inoperative, with its pin 4 output high, when its pin 1 input is high, but acts as a square-wave generator when pin 1 is low.

Thus when the pin 1 input terminal is high, zero base current is fed to Q_1, so the circuit is inoperative and passes only a small leakage current, but when the

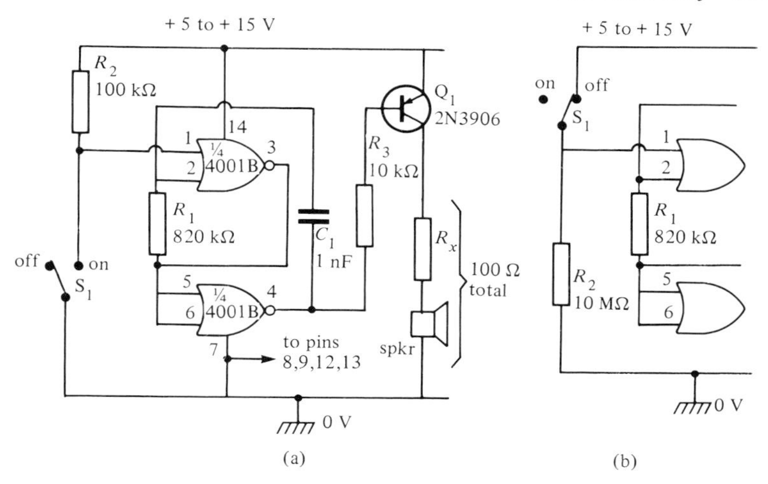

Figure 10.1 *Low-power 800 Hz monotone alarm-call generator designed for (a) NO and (b) NC switch activation*

input is low the astable is operative and generates a square-wave tone signal in the speaker via Q_1. Note that the circuit can be activated via normally open (NO) switch contacts by using the connections shown in *Figure 10.1(a)*, or by normally closed (NC) contacts by using the input connections shown in *Figure 10.1(b)*. In the latter case the circuit draws a standby current of about 1 μA via bias resistor R_2.

The basic *Figure 10.1* circuit is intended for low-power applications only, and can be used with any speaker in the range 3 Ω to 100 Ω and with any supply in the range 5 to 15 volts. Note that resistor R_x is wired in series with the speaker and must have its value chosen so that the total resistance is roughly 100 Ω, to keep the dissipation of Q_1 within acceptable limits. The actual power output level of the circuit depends on the individual values of speaker impedance and supply voltage that are used, but is no more than a few tens of milliwatts. If desired, the output power can be raised as high as 18 watts by using one of the booster circuits shown later in this chapter.

Figure 10.2 shows how a 4001B can be used to generate a pulsed 800 Hz alarm-call signal when its contacts are operated. Here, the two left-hand gates of the IC are wired as a low-frequency (about 6 Hz) gated astable multivibrator that is activated by the input switch, and the two right-hand gates are wired as an 800 Hz astable that is activated by the 6 Hz astable.

Normally, when the pin 1 input terminal is high, both astables are inoperative and the circuit consumes only a small leakage current. When the input is low, both astables are activated and the low-frequency circuit pulses the 800 Hz astable on and off at a rate of about 6 Hz, so a pulsed 800 Hz tone is

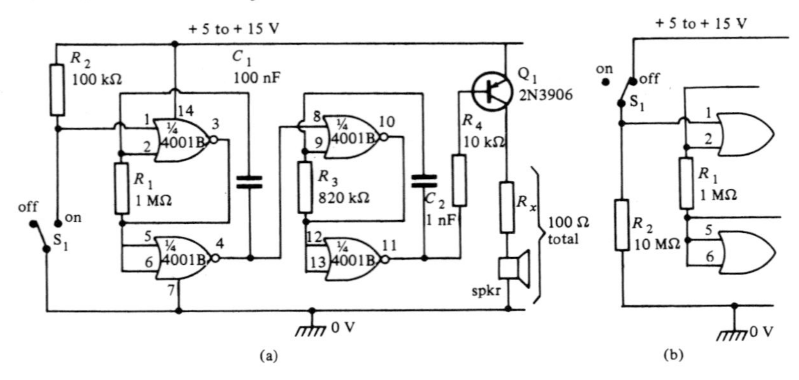

Figure 10.2 *Low-power pulsed-tone alarm-call generator designed for (a) NO and (b) NC switch activation*

generated in the speaker. Note that the circuit can be activated via an NO switch by using the connections shown in *Figure 10.2(a)* or by an NC switch by using the connections of *Figure 10.2(b)*.

Figure 10.3 shows how the *Figure 10.2* circuit can be modified so that it generates a warble-tone alarm signal that switches alternately between 600 Hz and 450 Hz at a 6 Hz rate. These two circuits are basically similar, but in *Figure 10.3* the 6 Hz astable is used to modulate the frequency of the right-hand astable rather than to pulse it on and off. Note that the pin 1 and pin 8 gate terminals of both astables are tied together, and the astables are thus both activated directly by the input switch. The circuit can be activated via NO switches by using the connections of *Figure 10.3(a)*, or by NC switches by using the connections of *Figure 10.3(b)*.

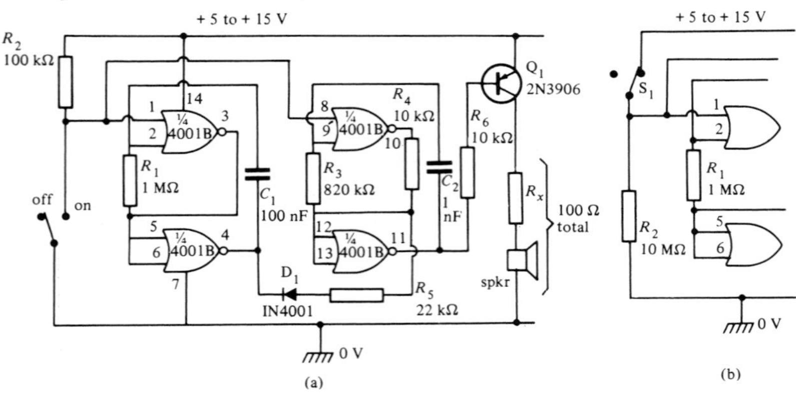

Figure 10.3 *Low-power warble-tone alarm-call generator designed for (a) NO and (b) NC switch activation*

Self-latching circuits

The alarm-call generator circuits of *Figures 10.1* to *10.3* are all non-latching types which produce an output only while activated by their control switches. By contrast, *Figures 10.4* and *10.5* show two ways of using a 4001B so that it gives some form of self-latching alarm-generating action.

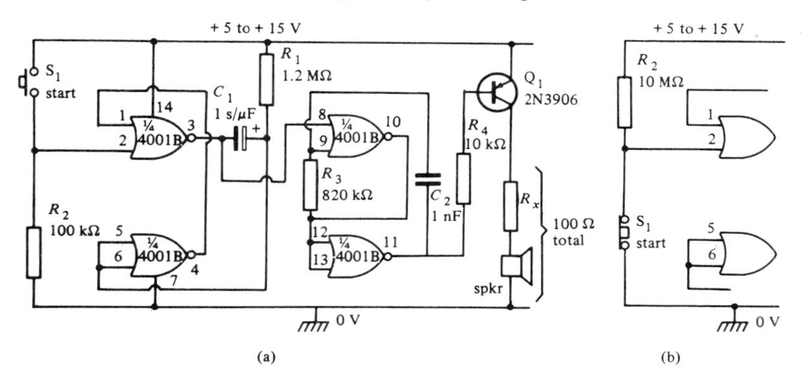

Figure 10.4 *One-shot 800 Hz monotone alarm designed for (a) NO and (b) NC switch activation*

The *Figure 10.4* circuit is that of a one-shot or auto-turn-off alarm-call generator. Here, the two left-hand gates of the IC are wired as a one-shot or monostable multivibrator that can be triggered by a rising voltage on pin 2, and the two right-hand gates are wired as a gated 800 Hz astable multivibrator that is activated by the output of the monostable.

Thus, the circuit action is such that both multivibrators are normally inoperative and the circuit consumes only a small leakage current. As soon as switch S_1 is momentarily activated, however, the monostable triggers and turns on the 800 Hz monotone alarm signal, which then continues to be generated for a preset period, irrespective of the state of S_1. At the end of this period the alarm automatically turns off again, and the action is complete. The circuit can be reactivated again by applying another rising voltage to pin 2 via S_1 (*Figures 10.4(a)* and *10.4(b)* show alternative switching methods). Note that the alarm duration time is determined by the C_1 value, and approximates one second per microfarad of value; periods of several minutes can readily be obtained.

Finally, *Figure 10.5* shows the circuit of a true self-latching 800 Hz switch-activated monotone alarm-call generator. Here, the two left-hand gates are wired as a manually triggered bistable multivibrator, and the two right-hand gates as a gated 800 Hz astable that is activated via the bistable.

The circuit action is such that the bistable output is normally high, so the

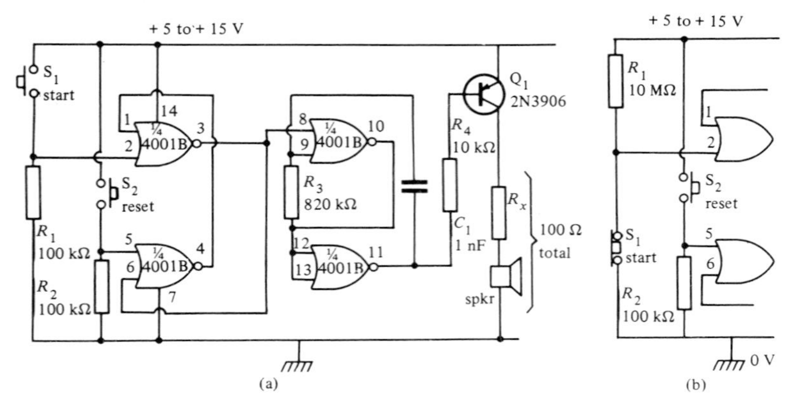

Figure 10.5 *Self-latching 800 Hz monotone alarm for (a) NO and (b) NC switch activation*

astable is disabled and the circuit consumes only a small leakage current. When S_1 is briefly operated a positive signal is fed to pin 2 of the IC, so the bistable changes state and its output locks low and activates the 800 Hz monotone generator. Once the alarm signal has been so activated, it can only be turned off again by removing the positive signal from pin 2 and briefly closing reset switch S_2, at which point the circuit resets and its quiescent current returns to leakage levels.

Power boosters

The mean output power of the *Figure 10.1* to *10.5* circuits depends on the individual values of speaker impedance and supply voltage used, but is usually of the order of only a few tens of milliwatts. Using a 9 V supply, for example, the output power to a 15 Ω speaker is about 25 mW, and to a 100 Ω speaker is about 160 mW. If desired, the output powers of these circuits can be greatly increased by modifying their outputs to accept the power booster circuits of *Figure 10.6* or *10.7*.

In these circuits, R_2 is wired in series with the collector of the existing Q_1 alarm output transistor and provides base drive to a one- or two-transistor booster stage, and the alarm's power supply is decoupled from that of the booster via R_1-C_1. Note that protection diodes are wired across the speakers of the booster circuits, and prevent the speaker back EMFs from exceeding the supply rail voltage.

The *Figure 10.6* booster circuit can be used with any speaker in the range 5 Ω to 25 Ω and with any supply from 5 V to 15 V. The available output power varies from 250 mW when a 25 Ω speaker is used with a 5 V supply, to 11.25 W when a 5 Ω speaker is used with a 15 V supply. The *Figure 10.7* circuit is

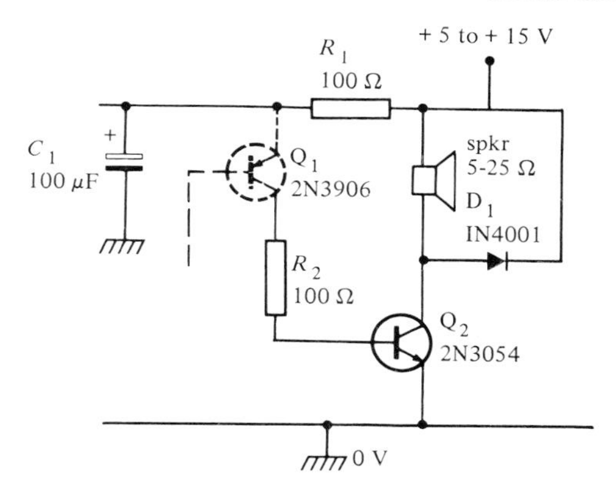

Figure 10.6 *Medium-power (0.25 W to 11.25 W) booster stage*

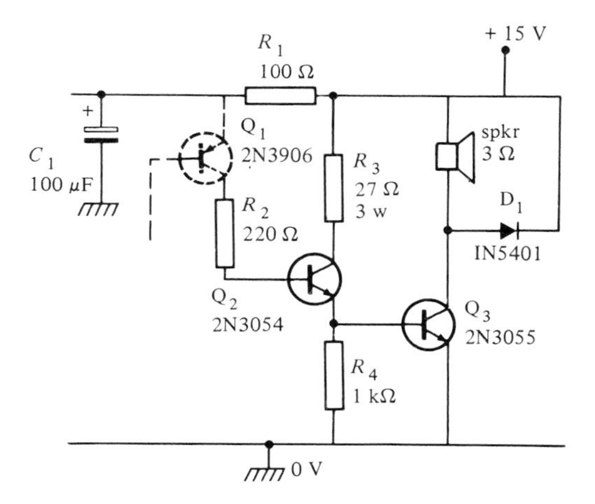

Figure 10.7 *High-power (18 W) booster stage*

designed to operate from a fixed 15 V supply and to use a 3 Ω speaker, and gives a mean output power of about 18 W. Note that, because of transistor leakage currents, the *Figure 10.6* and *10.7* circuits pass quiescent currents of roughly 10 μA and 30 μA, respectively, when in the standby mode.

Multitone alarms

Each of the *Figure 10.1* to *10.5* circuits has a single input switch and generates a distinctive sound when that switch is operated. By contrast, *Figures 10.8* and

10.9 show a couple of multitone alarm-call generators that each have two or three input switches and generate a distinctive sound via each input switch. These circuits are useful in identifier applications, such as in door announcing where, for example, a high tone may be generated via a front door switch, a low tone via a back door switch, and a medium tone via a side door switch.

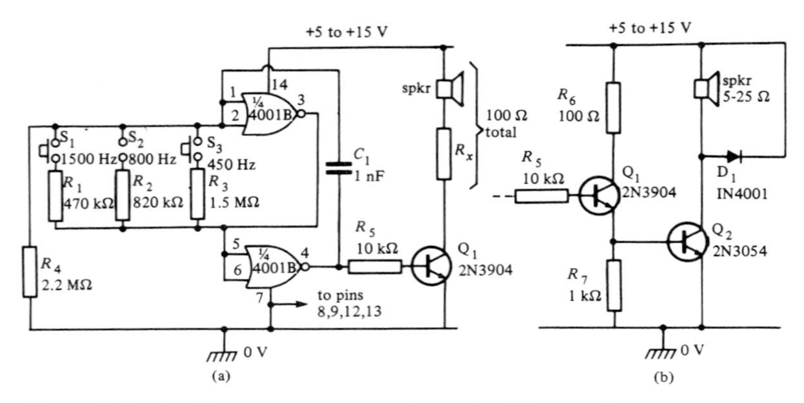

Figure 10.8 *(a) Three-input multitone alarm, plus (b) power booster stage*

The *Figure 10.8* circuit is that of a simple three-input monotone alarm-call generator. Here, two 4001B gates are wired as a modified astabled multivibrator. The action is such that the circuit is normally inoperative and drawing only a slight standby current, but becomes active and acts as a square-wave generator when a resistance is connected between pins 2 and 5 of the IC. This resistance must be less than the 2.2 MΩ value of R_4, and the frequency of the generated tone is inversely proportional to the resistance value that is used. With the component values shown, the circuit generates a tone of roughly 1500 Hz via S_1, 800 Hz via S_2, and 450 Hz via S_3. Note that these tones are each separated by about an octave, so each push-button generates a very distinctive tone.

As in the case of most other alarms already described, the *Figure 10.8 (a)* circuit generates an output power of only a few tens of milliwatts. If required, the output can be boosted as high as 11.25 W by using the booster stage of *Figure 10.8(b)*.

Figure 10.9 shows the circuit of a two-input multitone unit that generates a pulsed-tone signal via S_1 or a monotone signal via S_2. Here, the two left-hand gates of the IC are wired as a low-frequency (6 Hz) gated astable and the two right-hand ones as a 800 Hz astable. The two astables are interconnected via silicon diode D_1. The circuit action is such that the 6 Hz astable oscillates and activates the 800 Hz astable when S_1 is operated, thus producing a pulsed-tone output signal, but only the 800 Hz astable operates when S_2 is closed,

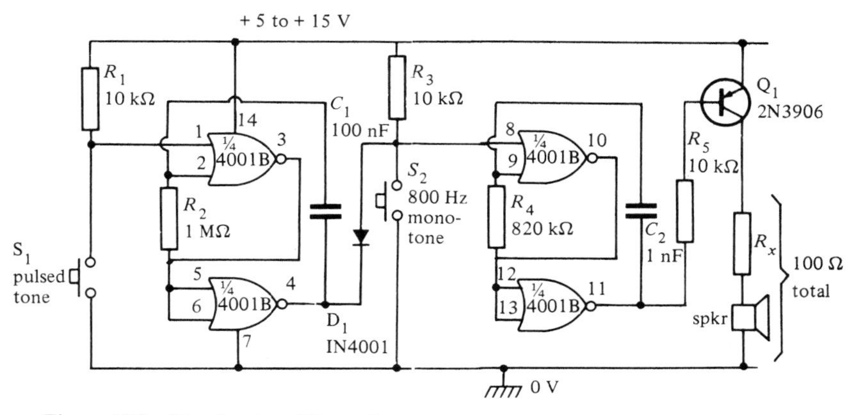

Figure 10.9 *Two-input multitone alarm*

thus producing a monotone output signal. This basic *Figure 10.9* circuit generates output powers of only a few tens of milliwatts, but if desired this power can be greatly boosted via the circuits of *Figure 10.6* or *10.7*.

Security alarms

Most security alarms are sound-generating systems that can be triggered by the opening or closing of one or more sets of electrical contacts. These contacts may be simple push-button switches, hidden pressure-pad switches, magnetically operated reed relays, etc., and the alarm systems may give an audible loudspeaker output, an alarm-bell output, or a relay output that can be used to operate any kind of audible alarm.

Security alarms have many uses in the home and in industry. They can be used to grab attention when someone operates a push switch, when an intruder opens a window or door or treads on a pressure-pad, when a piece of machinery moves beyond a preset limit and activates a microswitch, and so on. Several different types of relay-output security alarm are described in this section; the simplest of these is shown in *Figure 10.10*.

Here, S_1 is normally closed and relay A (RLA) is off, so the circuit consumes zero standby current. When any of the S_2 switches are briefly closed, the relay turns on and is self-latched via contacts RLA_2 and activates the alarm bell via contacts RLA_1. Once activated, the alarm can be reset by briefly opening S_1. Note that any number of NO switches can be wired in parallel in the S_2 position, and the alarm can thus be activated from any desired number of input points.

A weakness of the *Figure 10.10* design is that any one of its S_2 operating

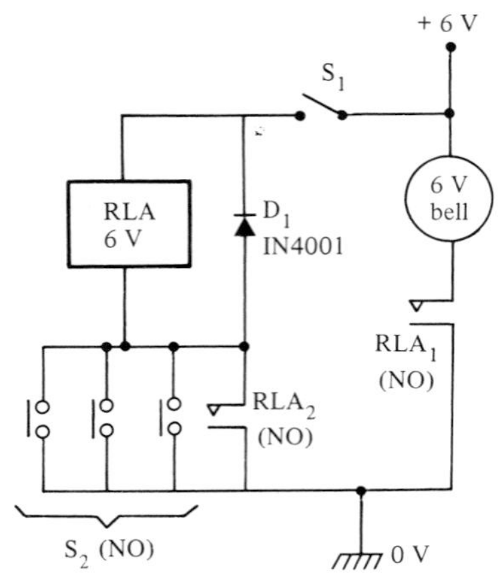

Figure 10.10 *Simple security alarm, activated by NO switches*

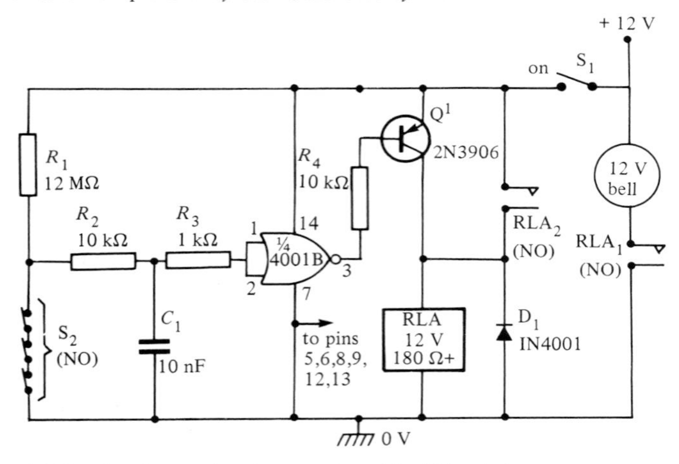

Figure 10.11 *Simple security alarm, activated by NC switches*

switches can be disabled by simply cutting through the cable that connects it to the main circuit. This snag can be overcome by using the design of *Figure 10.11*, in which all S_2 operating switches are NC types wired in series.

Here, the 4001B CMOS gate is wired as an inverter and activates the relay coil via Q_1. Normally, with all switches closed, the inverter output is high and Q_1 and the relay are off, and the circuit draws a quiescent current of about 1 μA via R_1. If any of the S_2 switches open or have their cables cut, the inverter output rapidly switches low and drives the relay on via Q_1, and the relay self-latches via contacts RLA_2 and activates the alarm bell via contacts RLA_1.

Note that R_2-C_1-R_3 acts as a 'lightning suppressor' network that filters out any rapid transient voltages that are induced into the S_2 switch cables by the action of lightning. These filter components should be placed as close to the inverter input as possible.

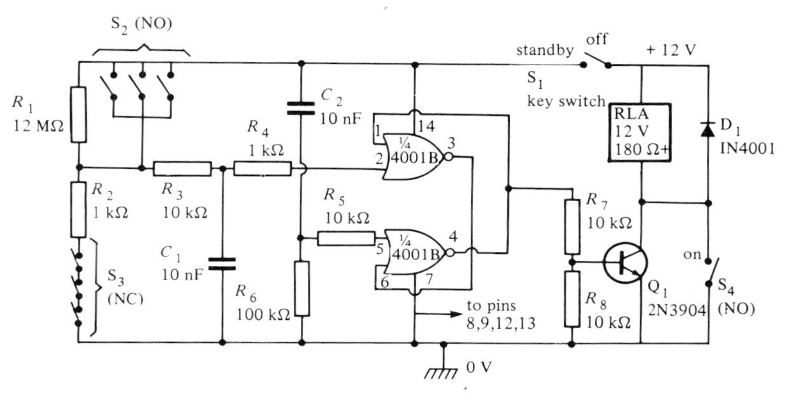

Figure 10.12 *Simple self-latching burglar alarm*

Figure 10.12 shows an example of a simple self-latching burglar alarm circuit that can be activated via any number of parallel-connected NO switches (S_2) or series-connected NC switches (S_3); the circuit can also be activated in the non-latching mode at any time via any number of parallel-connected NO switches (S_4).

In *Figure 10.12* the self-latching alarm action is provided by the two 4001B CMOS gates, which are wired as a bistable multivibrator and activate the relay via Q_1. Note that this bistable can be set (turn the relay on) via the S_2 or S_3 switches and the R_3-C_1-R_4 suppressor network, but is automatically reset via C_2-R_6 each time S_1 is switched to the standby position.

Figure 10.13 shows how the above circuit can be modified to give auto-turn-off (rather than self-latching) burglar alarm action, causing the relay to turn off after about four minutes of activation. In this case the two 4001B gates are simply wired in the monostable multivibrator mode, and the four-minute delay is determined by the C_3-R_9 values.

Next, *Figure 10.14* shows a utility burglar alarm that can be used in many domestic situations. It can be used with any number of series-connected NC sensor switches, and gives a self-latching alarm action via relay contacts RLA_1. The circuit action is such that an LED (driven via inverter IC_{1a} and emitter follower Q_1) illuminates if any of the NC sensor switches are open. However, the actual alarm system is automatically disabled (via the C_2-R_5 time-delay network) for an initial 50 seconds when key-switch S_3 is first set to standby, this giving the owner time to reset the switches or leave the house

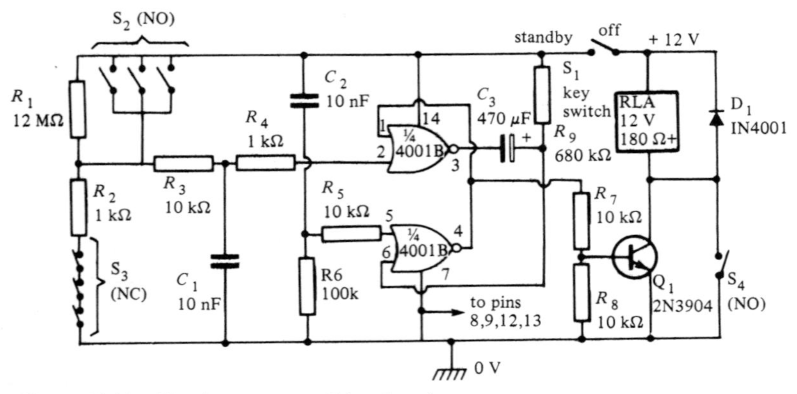

Figure 10.13 *Simple auto-turn-off burglar alarm*

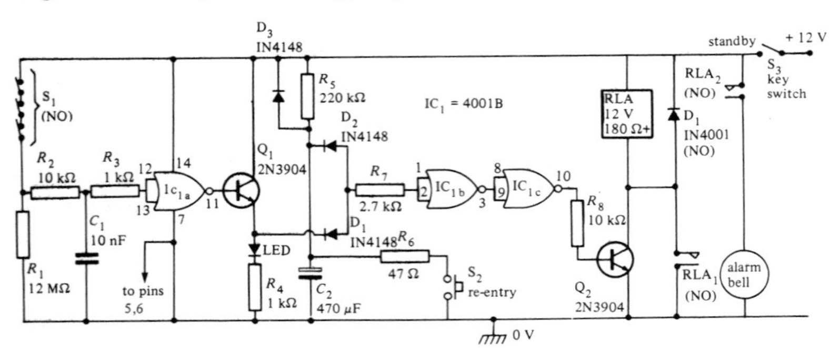

Figure 10.14 *Utility burglar alarm*

without sounding the alarm. The owner can subsequently re-enter the house without sounding the alarm by reactivating the 50 second delay via hidden external re-entry switch S_2. Note that the actual alarm bell (activated via relay contacts RLA_2) can use the same power supply as the *Figure 10.14* circuit if desired.

Finally, *Figure 10.15* shows how another relay (RLB) can be wired to the above circuit so that it also gives fire and panic protection. Here, NC push-button switch S_4 and NO switches S_5-S_6-RLB_1 are all wired in series with the coil of relay B, and the combination is permanently wired across the supply lines; the NO RLB_2 contacts are used to connect the existing alarm generator to the supply lines. Normally, RLB is off, but turns on and self-latches and activates the alarm generator if any of the S_5 or S_6 switches briefly close. Any number of NO panic buttons can be wired in parallel with S_5, and any number of NO thermostats can be wired in parallel with S_6. Once RLB has turned on, it can be turned off again via S_4, which may be in a hidden position.

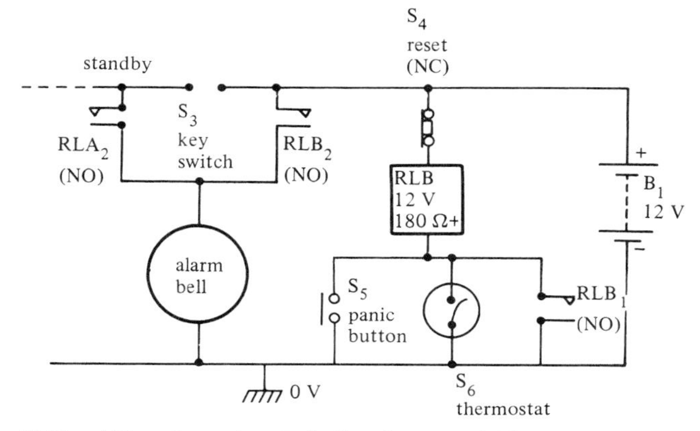

Figure 10.15 *Add-on fire and panic facility for use with* Figure 10.14

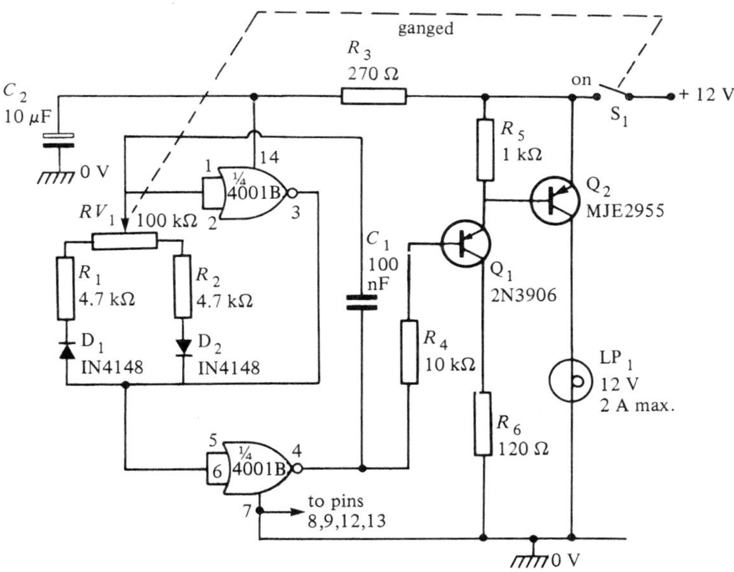

Figure 10.16 *12 volt DC lamp dimmer*

DC lamp dimmer

Figure 10.16 shows how a 4001B CMOS IC and a couple of transistors can be used to make a highly efficient lamp dimmer that can be used to control the brilliance of the internal lights of any car fitted with a 12 V negative-ground electrical system.

Here, the two CMOS gates are wired as a 100 Hz astable multivibrator that drives the lamp via Q_1-Q_2 and has its duty cycle or mark/space ratio fully variable from 1:20 to 20:1 via RV_1, thus enabling the mean power drive to the lamp to be varied from about 5 to 95 per cent of maximum via RV_1. Since the period of the 100 Hz waveform (10 ms) is short relative to the thermal time constant of the lamp, its brilliance can be varied from near-zero to maximum with no sign of flicker. Note that on/off switch S_1 is ganged to RV_1, so the circuit can be switched fully off by simply turning RV_1 fully anti-clockwise.

Manual clocking

In Chapter 7 we showed how simple clocked flip-flop ICs (such as the 4013B dual D-type) can be used as counter/dividers. One thing that we did not explain was how to clock these counters manually via a push-button switch. This omission is put right in *Figure 10.17*, which shows how to clock a single 4013B stage via S_1.

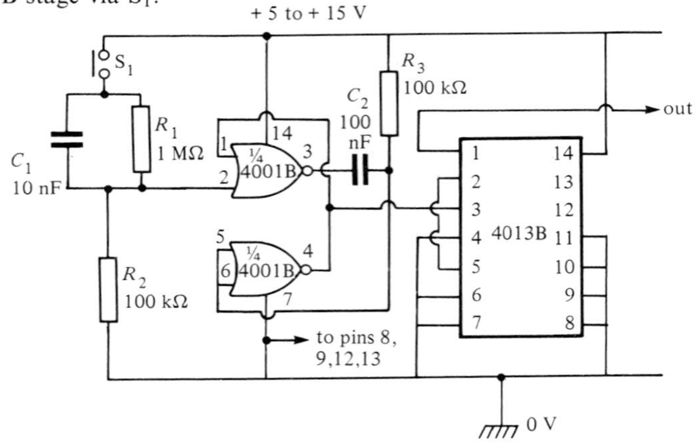

Figure 10.17 *Manual clocking of a 4013B counter/divider stage*

Here, the two 4001B gates are configured as a monostable multivibrator that generates a single 10 ms clock pulse each time S_1 is operated, thus causing the D-type flip-flop to change state; thus, S_1 must be operated twice to make the flip-flop go through a complete on/off or divide-by-two sequence. Note that the C_1-R_1 network effectively debounces S_1 and helps ensure clean triggering of the monostable circuit.

4046B PLL circuits

In Chapter 5 we introduced the 4046B phase-locked-loop (PLL) IC (see *Figure 5.25*) and showed how it can be used in voltage-controlled oscillator

(VCO) applications. This chip actually contains two phase comparators, a wide-range VCO, a zener diode, and a few other bits and pieces, and is specifically intended for use in PLL applications such as automatic frequency tracking, frequency multiplication, and frequency synthesis.

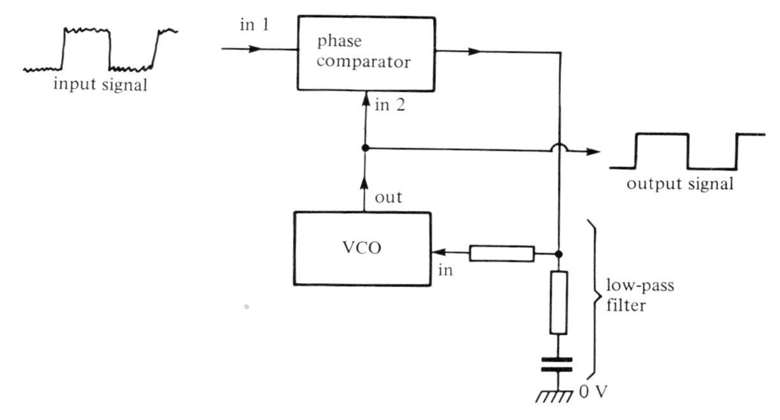

Figure 10.18 *Basic PLL frequency tracking circuit*

The basic operating principle of the PLL can be understood with the aid of the frequency tracking circuit of *Figure 10.18*. Here, the phase comparator element has two input terminals, one fed from an external input signal and the other from the output of the VCO element. The comparator compares the phase and frequency of the two inputs and generates an output proportional to their difference; this output signal is then smoothed via the low-pass filter network and fed to the control input terminal of the VCO, thus completing the phase-locked feedback loop.

The basic action of the above circuit is such that if the VCO frequency is below that of the external signal the comparator output goes positive and (via the filter network) causes the VCO frequency to increase until both its frequency and phase precisely match (phase lock with) those of the external signal. If the VCO frequency rises above that of the external signal the reverse action takes place, and the comparator output goes low and makes the VCO frequency decrease until it finally locks to that of the external signal. Thus, this circuit causes the VCO signal to automatically phase lock to the external input signal.

Note that in the above tracking circuit the VCO generates a clean (noise-free) and symmetrical output waveform, even if the input signal is noisy and non-symmetrical, and that (because the low-pass filter has a finite time constant) the VCO tracks the *mean* phase and frequency of the input signal. It can thus be used to track and clean up slowly varying input signals, or to track

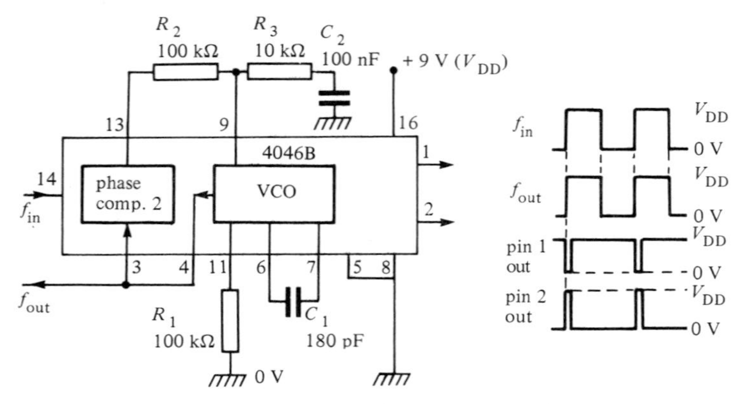

Figure 10.19 *Wide-range PLL signal tracker, showing waveforms obtained when the loop is locked*

the centre frequency of an FM signal and provide a demodulated signal at the comparator output.

Figure 10.19 shows how the 4046B can be used as a practical wide-range PLL that will capture and track any input signal within the 100 Hz to 100 kHz (approximate) span range of the VCO, provided that the pin 14 input signal switches fully between the logic-0 and logic-1 levels. Filter R_2-R_3-C_2 is used here as a sample-and-hold network, and its component values determine the settling and tracking times of signal capture. The VCO frequency is controlled by C_1-R_1 and the pin 9 voltage; the VCO span range (and thus the capture and tracking range of the circuit) varies from the frequency obtained with pin 9 at zero volts to that obtained with pin 9 at full supply rail value.

Figure 10.20 shows a lock detector/indicator that can be used with the above PLL circuit. The operating principle of this detector is moderately

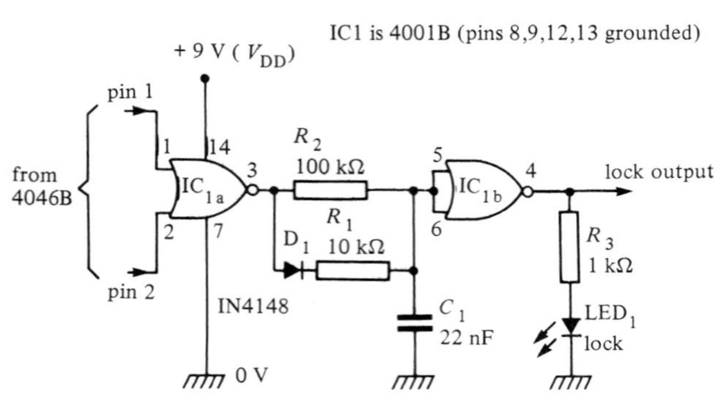

Figure 10.20 *PLL lock detector/indicator*

complex. Within the 4046B IC the output of each of the two phase comparators comprises a series of pulses with widths proportional to the difference between the two input signals of the comparator. When the PPL circuit is locked (see *Figure 10.19*) these two outputs are almost perfect mirror images of each other; when the loop is not locked the signals are greatly different. In *Figure 10.20* these two outputs are fed to the inputs of NOR gate IC_{1a}. The circuit action is such that when the loop is locked the IC_{1a} output is permanently low and illuminates LED_1 via IC_{1b}, but when the loop is not locked the output of IC_{1a} comprises a series of pulses that rapidly charge C_1 via D_1-R_1 and thus drive IC_{1b} output low and turn LED_1 off.

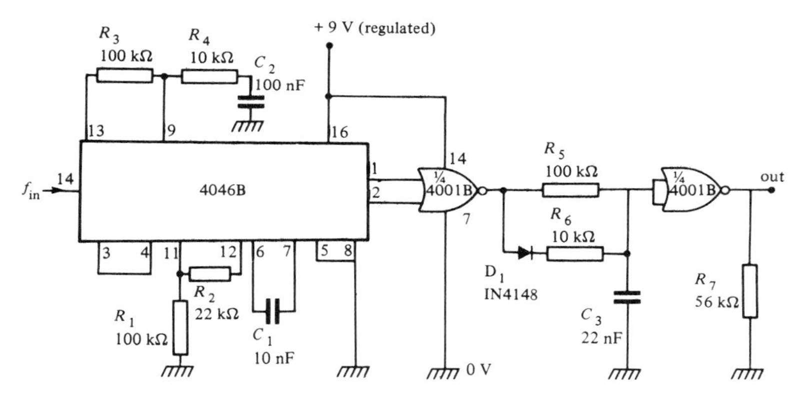

Figure 10.21 *Precision narrowband (1.8 kHz to 2.2 kHz) tone switch*

Figure 10.21 shows how a PLL circuit can be combined with the above lock indicator to make a precision narrowband tone switch. In this case the maximum VCO frequency is determined by C_1-R_1, and the minimum by C_1 and $R_1 + R_2$. The frequency is variable from about 1.8 kHz to 2.2 kHz with the component values shown, and the circuit can thus only lock to signals within this frequency range; the circuit output is normally low, but switches high when locked to a suitable input signal.

Frequency synthesis

One of the most useful applications of the PLL is as a frequency multiplier or synthesizer. *Figure 10.22* shows the basic principle. This circuit is similar to that of the basic PLL (*Figure 10.18*) circuit, except for the addition of the divide-by-*N* counter between the VCO output and the phase comparator input. The circuit action is such that the VCO frequency automatically locks to a value at which the divider output frequency matches that of the external

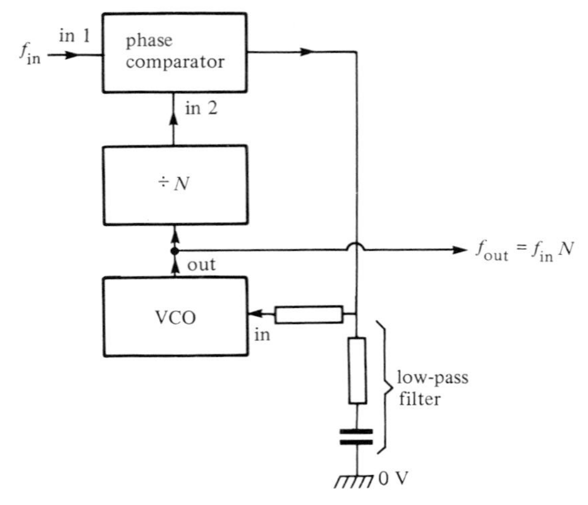

Figure 10.22 *Basic frequency synthesizer or multiplier circuit*

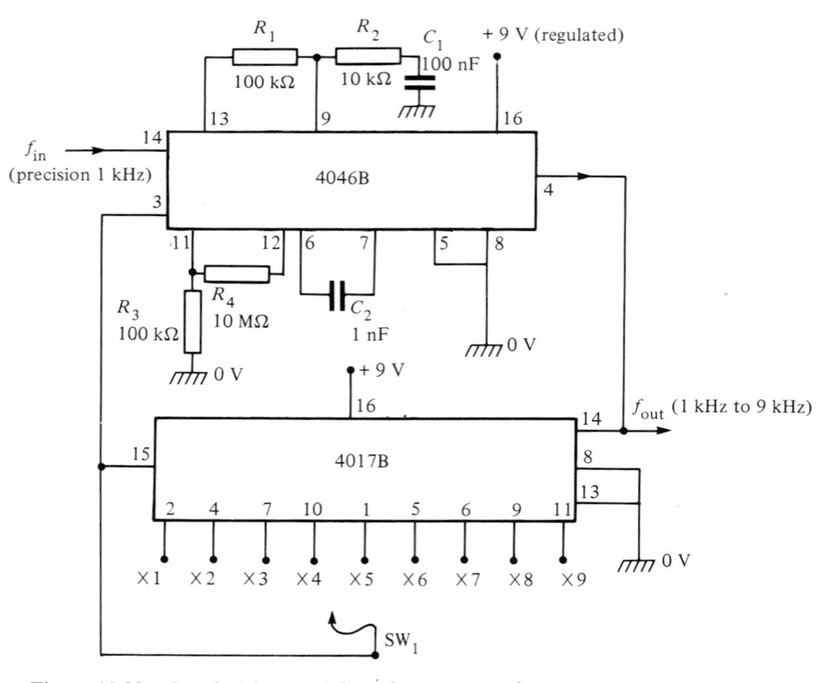

Figure 10.23 *Simple 1 kHz to 9 kHz frequency synthesizer*

input signal, and under this condition the VCO frequency is obviously N times the input frequency (where N is the counter's division ratio). If the input signal is derived from a precision crystal source, output signals of equal precision can thus be synthesized at any desired multiple frequency by simply using a divider with a suitable N value.

Figure 10.23 shows a practical example of a simple frequency synthesizer. It is fed with a precision (crystal-derived) 1 kHz input signal, and provides an output that is a whole-number multiple (in the range × 1 to × 9) of this signal. The 4017B is used as a programmable divide-by-N counter in this simple application, but can easily be replaced by a string of programmable decade down counters of the type described in Chapter 9, to make a wide-range (10 Hz to 1 MHz) synthesizer.

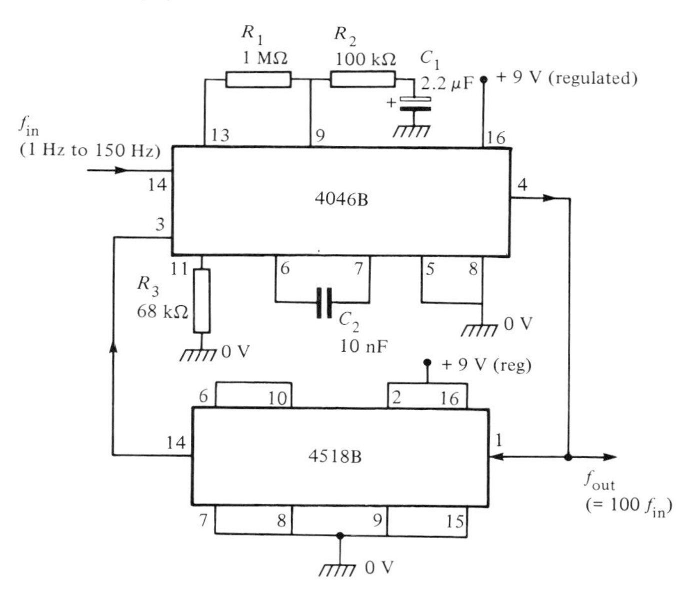

Figure 10.24 *A × 100 low-frequency prescaler*

Finally, to complete this volume, *Figure 10.24* shows how the synthesizer principle can be used to make a × 100 frequency prescaler that can be used to change a hard-to-measure 1 Hz to 150 Hz input signal into a 100 Hz to 15 kHz output signal that can easily be measured on a standard frequency counter. The 4518B IC used in this circuit actually contains a pair of decade counters, and in *Figure 10.24* these are cascaded to make a divide-by-100 counter.

Index

CMOS digital integrated circuits (by type number)